THE
CHEMISTRY OF LIFE

THE
CHEMISTRY OF LIFE

*Eight Lectures on
the History of Biochemistry*

by

ROBERT HILL   F. G. YOUNG

MALCOLM DIXON   LESLIE J. HARRIS

E. F. GALE   MIKULÁŠ TEICH

KENDAL DIXON   SIR RUDOLPH PETERS

Edited, with an Introduction, by

JOSEPH NEEDHAM

CAMBRIDGE

AT THE UNIVERSITY PRESS

1970

Published by the Syndics of the Cambridge University Press
Bentley House, 200 Euston Road, London N.W.1
American Branch: 32 East 57th Street, New York, N.Y.10022

© Cambridge University Press 1970

Library of Congress Catalogue Card Number: 78–85733

Standard Book Number: 521 07379 0

Printed in Great Britain
at the University Printing House, Cambridge
(Brooke Crutchley, University Printer)

CONTENTS

v

CONTRIBUTORS

Kendal C. Dixon, M.D., *Reader in Cytopathology, University of Cambridge; Fellow of King's College*

Malcolm Dixon, F.R.S., *Emeritus Professor of Enzyme Biochemistry, University of Cambridge; Honorary Fellow of King's College*

E. F. Gale, F.R.S., *Professor of Chemical Microbiology, University of Cambridge; Fellow of St John's College*

Leslie J. Harris, SC.D., *formerly Director of the Dunn Nutritional Laboratory, University of Cambridge and Medical Research Council*

R. Hill, F.R.S., *formerly a member of the Scientific Staff of the Agricultural Research Council; Honorary Fellow of Emmanuel College*

Joseph Needham, F.R.S., *Master of Gonville and Caius College, Cambridge; Emeritus Sir William Dunn Reader in Biochemistry*

Sir Rudolph Peters, M.D., F.R.S., *Emeritus Whitley Professor of Biochemistry, University of Oxford; Honorary Fellow of Gonville & Caius College*

Mikuláš Teich, *of the Department of the History of Science and Technology, Institute of History, Czechoslovak Academy of Sciences, Prague; Visiting Scholar at King's College*

F. G. Young, F.R.S., *Sir William Dunn Professor of Biochemistry, University of Cambridge; Master of Darwin College*

INTRODUCTION

by Joseph Needham

This book, the outcome of a course of lectures which was given·
by Cambridge biochemists each year from 1958 to 1961 under
the aegis of the History of Science Committee and Department,
is not the first collective work which has been generated by the
modern movement for the history of science in Cambridge. For
in 1938 the University Press published *Background to modern
science*, ten lectures by leading figures in research who were
willing to look back over the previous four or five decades and
to say how the development of science seemed to them to have
been set forward during that time in their own field. The
modern phase of the history of science in Cambridge goes back
close on thirty-five years, for it was in 1936 that two events took
place: the loan exhibition of historic scientific apparatus in the
Old Schools organized by the late Hamshaw Thomas; and the
formation of a Committee (under the chairmanship of the present editor and with Walter Pagel as secretary) to arrange for
courses of lectures on the history of science. From these small
beginnings over the years has grown the present thriving Department of the History and Philosophy of Science (headed by
Gerd Buchdahl), with its Whipple Museum of historical scientific
instruments and a valuable library used by many students and
research workers. The lectures now printed in this book were
well attended, and the group was happy to welcome into it one
player from away, namely Mikuláš Teich of Prague, one of the
few living professional historians of biochemistry. Although our
subject is of such fascinating borderline interest, it has so far
surprisingly given rise to only one full-scale book, that of Fritz
Lieben (1935); perhaps Teich will endow us with another.

Although some far backward glances are here included,
effectively most of the material in this book refers to events from
about 1800 onwards. If we are to talk about the history of
biochemistry, the development of human knowledge of the

chemistry of life, we are faced with the problem of defining it. It is quite a tenable point of view to hold that it is not possible to speak of biochemistry until modern chemistry had been born —and that did not happen until the growth of the chemistry of gases in the latter half of the eighteenth century, the 'revolution in chemistry' of Lavoisier, and the founding of modern atomic chemistry by Dalton. But this may well be thought too narrow and rigid, laying the accent also on the '-chemistry' rather than the 'bio-'. By contrast, if one includes all the old ideas, speculations and groping experiments on the chemistry of living things, whether plants or animals, then a vast realm of history lies before us. Nothing that men have ever thought and done on the implicit presupposition that there *was* a chemistry of living matter, can then be alien to us. Some selection from the many trends of thought involved was what I used to contribute in my lecture in the series now before us, but there is such a gulf fixed between ancient and mediaeval ideas and doings on the one hand, and modern post-Renaissance scientific endeavour on the other, that my participation here may better take the form of an introduction to the whole, even though it means compressing a quart into a pint pot. What is needed is some sketch of the background panorama for the contributions which follow, a perspective glass to show that their subjects did not suddenly come into being like a set of extraordinary mutations, but rather that they grew like plants with roots coming up a very long way from the sub-soil of antiquity. Furthermore, it will be found when one penetrates into the bygone origins of biochemical ideas that it is necessary to take the whole of the Old World into consideration, not treating Europe as a self-contained entity, which it never was, but bringing into consideration the great contributions of the Arabic, Indian and Chinese culture-areas.

In the very restricted space at my disposal it will be possible neither to give illustrative quotations, always so moving in the words of the original writer, nor to give precise datings except for names likely to be unfamiliar to the general reader, nor to give chapter and verse for statements that will be made. A select group of references may, I hope, atone somewhat for this. How best in short space to depict that general background already

Introduction

mentioned presents, of course, an almost insoluble problem. One could search the pre-Socratic philosophers (sixth and fifth centuries B.C.) and the Hippocratic corpus (fourth century B.C.) for many interesting speculations about the material basis of life, and to them one could add similar material from the *Suśruta* and *Caraka Samhitas* in India (first to third centuries A.D., revised eighth to eleventh) and from the philosophers of the Warring States period (fifth to third centuries B.C.) further east. Alternatively, one could take a single conception, for example animal heat, and follow it through all its ramifications from the earliest Greek and Chinese statements down to the beginnings of modern knowledge of oxidation-reduction processes in the living cell. I thought however that I should like to follow a different way and dwell less extensively on a small number of key ideas, almost as if they were entries in a kind of Voltairean 'Dictionnaire Philosophique' (without however any undertone of mockery). Let us therefore try to say something about each of the following key-words—Pneuma—Element—Humour—Krasis—Quintessence—Elixir—Conjunction—Ferment.

How many books could be written (and of course have been written) on the 'breath of life'! The association of breathing with life and its cessation with death came down from the earliest historical times and beyond, and it was quite understandable that the functions of the body and all its organs and tissues should be thought of in terms of the movement and nature of a number of imperceptible breaths. This pneumatic proto-physiology and proto-biochemistry was common to all the civilizations of the Old World. For it has been shown in detail that the *pneuma* of the Greeks was equivalent to the *prana* of India, the *ch'i* of the Chinese and later on the *rūḥ* of the Arabs. The universality of this conception prompts the conviction that it originated in the Fertile Crescent, spreading out in all directions, even if so far the Assyriologists have not been able to find much evidence for this. The Indians classically counted and tried to define five sorts of *prana* and the Chinese a good many more. For a millennium and a half European thought on these matters was dominated by the Galenic theory of spirits; *pneuma,* spirit as such, coming from the circumambient air through the lungs, and combining with *pneuma physikon* or natural spirits prepared by the action of the liver on the food, to form *pneuma zōtikon* or vital spirits. The former were distributed in ebb and flow from the right ventricle of the heart through the system of the veins, the latter from the left ventricle through the arteries. Similarly, under the action of the brain, the vital spirits were transformed into *pneuma psychikon* or animal spirits, distributed to all parts of the body through the nerves. Here there was of course some connexion with the Aristotelian doctrine of 'souls', for both the natural and vital spirits were at the level of the vegetal soul (*psychē threptikē*), and the animal spirits were in the domain of the sensitive soul (*psychē aisthētikē*), while for the rational soul (*psychē dianoētikē*), because essentially mental, there was no equivalent in the world of material spirits. Ancient Chinese ideas were much more like those of the Greeks than has usually been supposed, for at about the same time as Aristotle, Hsün Ch'ing and others conceived of a 'ladder of souls', e.g. *ch'i*, *shêng*, *chih* and *i*, representing a very similar ascent. From our present viewpoint, the 'souls' were but names for particular functions of the levels of living, yet we still conserve such names as 'vegetal' for the yolky poles of eggs, and 'vegetative' for parts of the nervous system.

When the Renaissance came, something like nitre and salt-petre was recognized in the external *pneuma*, and the 'nitro-aerial particles', surmised by the Paracelsians and demon-strated by Mayow, pointed the way towards the development of 'pneumatic' chemistry in the eighteenth century. Meanwhile John-Baptist van Helmont (1579 to 1644), who has perhaps as good a title as anyone else to be called the father and founder of all modern biochemistry, separated clearly the ancient *pneuma* idea into his *gas* and his *blas*. From his recognition of the identity of carbon dioxide onwards, the way led clear through Priestley, Cavendish, Black and all the others to the modern discoveries of the transit and carriage of gases in living bodies, the respiration of the cells themselves, and the nature of the energy-providing oxidation-reduction reactions going on inside them. Van Helmont's *blas* was another way of talking about the more spiritual spirits, including those *archaei* which both

Introduction

Paracelsians and Chinese had visualized as presiding over the activities of every organ (like controllers sitting in front of instrument-panels in glass boxes high above the automatized activities of great factory halls in modern industrial plants). He thus skilfully shunted off all discussions of this aspect of things on to the philosophical loop-line of the mechanist-vitalist controversies. These have been prominent at various times, as in the Naturphilosophie period early in the nineteenth century, the time of the synthesis of urea by Wöhler (lecture 7 hereafter), or again in the Liebig–Pasteur controversy (lecture 3), or again in the present century, aroused by the supposed gas-secretory activity of the lung; but today they have lost much of their intensity with the recognition of the omnipresence of levels of organization, morphology extending down to the level of complex molecules, and chemical processes reaching up to the very substratum of mental activity (lecture 4).

In China also, *ch'i* became more and more material as time went on, and having begun (as we know from the etymology of its ideograph) as the steam arising from cooking rice, it came to mean by the twelfth century all kinds of matter, however gross. This was the time at which the Chinese scholastics, the Neo-Confucians, were able to build their scientific world-view on two conceptions only, *ch'i*, which we might now call matter-energy, and *li*, which we may call organic pattern at all levels, wherever manifested. The pursuit of the *ch'i* led people in China to some striking discoveries, such as that of the existence of deficiency diseases (cf. lecture 6) by a Mongolian physician about A.D. 1325. This was Hu-Ssu-Hui (Hoshoi), Imperial Dietician under four Yuan emperors, who observed that there were some diseases, such as, for example, the last stages of avitaminosis-B, which could be cured by diet alone, without the intervention of any drugs at all. We still possess his book, the *Yin Shan Chêng Yao*.

All this pneumatic heritage has left indelible traces on both our language and our thinking. As I sat down to write these words, builder's tradesmen were loudly singing 'in high spirits' on the other side of the garden wall, and in the newspaper I had read of the evidence transmitted back by Mariner 7 that there is methane and ammonia—therefore perhaps *life*—on the planet Mars.

Introduction

For two thousand years or more before the appearance of the modern conception of elementary substance, during and after the Lavoisierian revolution, all the peoples of the Old World had had the conception of Elements. The classical Aristotelian earth, fire, air and water are well known, and could still inspire modern poets such as W. B. Yeats and Dorothy Wellesley; less well known are the Chinese five: metal, wood, water, fire and earth. These played perhaps a greater role in proto-physiology and proto-biochemistry at the eastern end of the Old World than the four elements did in the West, for in accordance with the elaborate Chinese system of symbolic correlations, elements were associated not only with planets but with specific organs in the body. That they could conquer each other and be generated from one another was again an idea which existed among the Greeks (as in Plato's *Timaeus*) but which was carried much further in Chinese thought. Earth, fire, air and water came right down to the beginnings of modern chemistry, for as late as the last quarter of the seventeenth century it was a great point of dispute (in which Robert Boyle prominently participated) whether or not analysis of mixed bodies by fire could demonstrate either the existence of the classical elements or that of the *tria prima*, salt, sulphur and mercury, which the Paracelsians (cf. p. xx) had partially substituted for them. The Indian Buddhists also had their elements, as usual more like the Greeks with four, but India was particularly characterized by the development of a subtle and elaborate theoretical atomic system, especially in the Vaiseśika school. A good deal of work has been done on this, notably by philosophers, but it has not so far been adequately investigated by historians of science.

Parallel with the elements went the doctrine of the Humours, rounded out (though not very consistently) in the Hippocratic corpus; four fluids, as we should say, with different biochemical characteristics. Broadly speaking, the humours were first, blood, hot and wet (corresponding to air); secondly yellow bile or *chole*, hot and dry (corresponding to fire); thirdly black bile or *melanchole*, something like serum and perhaps derived from the observation of the lower part of a blood clot, cold and dry (corresponding to earth); and fourthly phlegm, cold and wet

(corresponding to water). Later on these humours were connected with the liver, gall-bladder, spleen and lungs respectively. Health depended, when not disturbed by trauma or by some invading influences, on the right balance (*krasis*) between the four humours. Such conceptions were actively employed in Western medicine down to the middle of the eighteenth century. In China and India similar note was taken of the characteristic juices and fluids of the body, but the emphasis was different, saliva and semen being more important than the biles. In the early Middle Ages the Taoists developed a particularly interesting theory of the three primary vitalities (*san yuan*), i.e. pneumatic essence (*ch'i*), seminal essence (*ching*), and mental essence (*shen*). These ideas seem to have come through to Europe in the late eighteenth century, and to have affected a number of biological thinkers in the Naturphilosophie period, who distinguished in various ways a tripartite set of bodily functions such as locomotion, secretion-generation and intellection. As for *krasis*, the Chinese made even more widespread use of the idea than the Greeks, for in their natural philosophy two fundamental forces, the Yin and the Yang (pairs of opposites, as darkness to light, female to male, cold and wet to hot and dry, etc.) informed the entire universe in every particular. It was most natural, therefore, that an imbalance between these two fundamental forces in the body would lead to disease, and century after century the physicians of China laboured to restore the natural and perfect equilibrium. No one, I am sure, among the contributors to this volume, inheritors all of Frederick Gowland Hopkins' famous epigram that 'Life is a dynamic equilibrium in a polyphasic system', would want to take exception to the insights of their Greek and Chinese ancestors. An inhibition in the activity of an enzyme or the supply of a substrate (cf. lecture 2) will certainly throw a spanner in the works of that extremely delicate balance on which the health of the cell depends; indeed also the normal activity of the mind itself (lecture 4); and it is needless to say that without the perfect synchronization of the endocrine orchestra (lecture 5), that marvellous symphony performed with no conductor, a hundred things will go painfully wrong with the bodies of men and

beasts. We may not use the expressions *krasis* and *t'iao ho* much today, but we still have to think in those terms.

Now for the Quintessence, an idea associated with the technique of distillation which played so great a part in both Arabic and European alchemy from the late ninth century to the beginning of the iatro-chemical period in the sixteenth. Here was something over and above the four elements, not always thought of strictly as a material substance, but more like *rūḥ* or *pneuma*— 'a volatile principle that could be separated from a material substance and possessed its characteristic activity in the highest degree'. Thus it was an especially subtle form of matter, intermediate almost between matter and spirit. This may be said to have been the beginning of the conception of 'concentrating' active principles, one of the basic methods of all modern biochemistry (cf. lectures 2, 5, 6). Organic substances had been subjected to distillation either wet or dry since the time of the Alexandrian proto-chemists onwards (first century A.D.). Since eggs were such obvious sources of potential life-force, the yolk and white were often distilled separately or together, giving the Alexandrians various fractions containing sulphur (the *theion hydōr*, 'divine' or 'sulphur' water) which had startling effects on the surfaces of metal alloys. The making and employment of such calcium polysulphides was one of the most interesting of all the Alexandrian procedures analogous to the preparation of stannic sulphide ('mosaic gold') by the Chinese.

Then someone distilled wine, and things began to happen. The discovery of this quintessence, alcohol, has been placed in Italy in the twelfth or even the eleventh century, but it is more likely to have been developed in the Arabic culture-area where al-Razī was distilling vinegar in the tenth and al-Kindī essential oils of perfumes in the ninth. It may even have started in China, where there are hitherto unexplained T'ang references (seventh and eighth centuries) to 'burnt wine' (cf. *branntwein* and brandy). The Jābirian corpus (about 900 A.D.), of which I shall say more in a moment, has a special 'Book of the Fifth Nature', and elsewhere there is a statement that the distillate of wine catches fire. Nevertheless, the Arab pharmacists seem not to have used *aqua ardens* as a drug, and it caused little stir until the

Westerners began to distil it themselves. It is very prominent in the Lullian corpus (after 1350, cf. p. xix), and the link with alchemy, gold and immortality came in because alcohol was soon found to have the property of making organic bodies 'incorruptible'. The complex liqueurs made by members of the religious orders (such as Benedictine, invented in 1510), with their great variety of plant ingredients, are actual examples of a microcosmic 'philosophical heaven', in which the stellar influences generating the virtues of many plants have been, as it were, captured on earth and 'fixed' for human benefit. From this point it was but a short step to the first use of alcohol as a fixative by Robert Boyle, who in 1666 described 'a way of preserving birds taken out of the egge, and other small faetuses'. As a name and a concept Quintessence died, but the hundreds and thousands of biologically active substances, of pigments, enzymes, hormones, vitamins, and other chemical molecules extracted with infinite labour from large amounts of starting material, are all its lineal descendants. Also we still buy 'eau de vie', and are still asked if we 'have any spirits to declare'.

A quintessence could also be an Elixir—and with this evocative word we are plunged at once into that ocean of words which have been written by and about alchemists since the beginning of our era. This is one of the most complex and difficult subjects of all historical research, but in laying bare the roots of modern biochemistry it cannot be overlooked. Almost from the beginning of historical time, with the life-giving potion of the Indian Vedic literature, *soma*, men have sought for some compounded herbal, fungal, mineral or metallic drug or tincture which would keep that breath of life gently pulsating long after the individual's 'appointed span'. Let no one say that this was an impossible dream, for today when antibiotics and sulpha-drugs have gained the victory over so many diseases formerly rabid killers, we are now facing the social problems not only of incurable diseases prevented in former times from manifesting themselves, but also of what to do with the generations that medical science preserves beyond their time. The word elixir, in the opinion of many, serves well to define alchemy itself, for in its earliest Western forms proto-chemistry was a matter of

making various kinds of artificial 'gold', and had litt
nothing of the 'macrobiotic' or longevity preoccupation.
came into Europe only with the transmission of Arabic alche
and since alchemy is after all itself an Arabic word, we sho
not speak of it in Europe until after the translations of the n
twelfth century. It then took some time to exert its full effe
but the emphasis on a longevity which chemistry could produ
reached full force in the writings of Roger Bacon (1214–92).

The traditional origin of the word elixir was from the Gree
xerion, a medicinal powder strewn upon wounds, or the powde
of projection which turned the base metal into gold; but now
seems at least as likely that the word is of Chinese provenance
perhaps from *yao chi* (medicinal dose), perhaps from *i chih*
(secreted juice). The first words in both these phrases ended
with a *k* in old Chinese pronunciation, which would account for
the *q* in *iqsir*. However this may be, it is certain that Chinese
proto-chemistry was by far the oldest to emphasize consistently
the macrobiotic preoccupation; and for the reason that in China
alone was there the idea of 'material immortality', the ethereal-
ization of the body and spirit so that the individual could
continue on earth to enjoy the beauty of Nature without end,
uncramped by the needs and limitations of ordinary mortals.
Here two points need emphasis: first, the association between
the manufacture of the imperishable metal, gold, and the attain-
ment by man of earthly imperishability begins with this earliest
alchemy (strictly so-called) of which we have any knowledge,
namely, that of China in the second century B.C.; secondly, the
thought-connexion between gold and immortality through the
centuries was that all the other metals, rusting and corroding,
suffered from the same illness as mortal man, so that the
'philosopher's stone', as it came to be called, was the supreme
medicine of man as well as metals.

If we were to draw a thumbnail sketch of the pre-natal
history of chemistry in the Old World, the summary would be
something like this. One has to distinguish half a dozen spatio-
temporal entities and phases. Beginning nearest home, there
was the Hellenistic or Graeco-Egyptian tradition of proto-
chemistry, starting in the first century A.D., continuing through

the Byzantine culture, and ending in the eleventh century, from which time the oldest extant MS texts have come down to us. This at least is the case for the philosophical writings, but we have much earlier MSS for the purely practical chemical-metallurgical techniques (e.g. of imitating gold), which centre on papyri of the third century A.D. This practical tradition continued on without much break till the eleventh century, through the *Compositiones ad Tingenda*, translated into Latin about 750, and the technical book of Theophilus Presbyter just before 1000. None of this was alchemy in the full sense, because it lacked the macrobiotic component; people were either imitating gold or believing that they had made an artificial form of it.

Far away at the other end of the Old World, however, there was another tradition in which the making of elixirs, or 'biochemical' medicines of immortality, was closely tied to artificial gold-making. This was the alchemy of China, ancestor of all the rest; it generated the first book on the subject in any civilization, the *Ts'an T'ung Ch'i* of Wei Po-Yang in A.D. 142, and later on, in 848, the first printed book on a chemical subject in any civilization, the *Hsüan Chieh Lu* of Hokan Chi on plant-drug antidotes for elixir-poisoning by heavy metals. So far there has been very little appreciation of the series of priorities and parallelisms between Chinese alchemy and Hellenistic-Byzantine proto-chemistry. The philosopher Tsou Yen (*fl.* 300 B.C.) precedes Bolus of Mendes the Democritean (*fl.* 200 B.C.) just as the metallurgical chemistry of the *K'ao Kung Chi* in the third century B.C. precedes the practical papyri of the third century A.D. In other ways one finds close parallelisms. The first clear connexion between gold and elixirs occurs in the statement of Li Shao-Chün (133 B.C.), but in the middle of the next century Anaxilaus of Larissa and Liu Hsiang were both active in attempting to make 'gold'. The first century A.D. brings the writings of Pseudo-Democritus and his group (Mary the Jewess, Pseudo-Cleopatra, Comarius, etc.) but although most valuable for the origins of chemical apparatus and techniques, they were basically inorganic in tendency and the *Physika kai Mystika* is not an essentially elixir book like that of Wei Po-Yang. Exactly the same contrast holds good of the two great systematizers writing about A.D. 300, Zosimus of Panopolis in Egypt, and Ko Hung (Pao P'u Tzu) in China; and it can be illustrated again by the men working some two or three centuries later, Olympiodorus the Neo-Platonist with Stephanus of Alexandria, on the one hand, as against the great alchemical physicians, T'ao Hung-Ching and Sun Ssu-Mo. Finally there is the strange coincidence that just as our codices of Hellenistic proto-chemical material date from about A.D. 1000, so the first definitive collection of the *Tao Tsang* (Taoist Patrology), home of so many alchemical books, took place in 1019, and the first printing of it in 1117.

Nevertheless, the Golden Age of Chinese alchemy had corresponded with the period of the T'ang dynasty (620–900 approximately). This had been the time of activity of men such as Mêng Shen, Mei Piao and Chao Nai-An. Since it is now quite clear that the alchemy of the Arabic culture-area was powerfully influenced from China, this lends additional significance to the fact that it flourished not as used to be thought in the seventh and eighth centuries, but rather in the ninth, tenth and eleventh. There was a veritable burst of activity around the year 900. The many writings of the Jābirian corpus were approximately contemporary with the lifetime of the great alchemist-physician al-Razī (860–925), but they were quite independent of him and related rather to the Qarmatian scientific group known as the Brethren of Sincerity (Ikhwan al-Ṣafā). This was also the time of the mystical alchemy of Ibn 'Umail and the compilation of the work known to the Latins as *Turba Philosophorum*—a kind of imaginary international chemical congress embodying lively statements of opinions by all the natural philosophers whom the Arabs knew of from the pre-Socratics downwards. A century later came the activity of the great Ibn Sīnā and al-Majritī, so that we are brought almost to the moment when the curtain went up on true alchemy in the Frankish or Latin West.

The first translations from Arabic into Latin were made in Spain, where al-Majritī had worked, from 1130 onwards, and for a whole hundred years after that an abundance of translation continued. Europeans were now excited by the idea of a biochemical medicine of longevity as well as the artificial

Introduction

making of gold; so that the considerable advances in chemical knowledge recorded make the fourteenth century a notable period. It begins with the books attributed to Geber, now recognized as original, not translations, and having nothing but the name in common with the earlier corpus of writings attributed to Jābir ibn Ḥayyān. The unknown author who covered himself with the Latin form of the name was nevertheless well acquainted with Arabic alchemy. Then followed Petrus Bonus (*fl.* 1330) and John of Rupescissa (*fl.* 1345), as well as two important collections of related MSS: the Villanovan corpus called after Arnold of Villanova and dating from the early fourteenth century, and the Lullian corpus called after Raymund Lull, from the late fourteenth. The fifteenth century was a less original time of stocktaking and of compilation, but the beginning of the sixteenth century saw an entirely new period come to birth.

The 'scientific revolution' is generally made to centre around the figures of Copernicus, Kepler and Galileo with all that they did towards the mathematization of natural phenomena, but there are many ways in which the work of Paracelsus (1493 to 1541) was hardly less revolutionary. He it was who broke the link between 'gold-making' and macrobiotics which Li Shao-Chün and his companions had forged in the second century B.C.; this he did with his famous epigram that 'The business of alchemy is not to make gold but to make medicines'. In the year before the death of Paracelsus was born the first of the modern chemists, Andreas Libavius, whose *Alchemia* of 1597 sought to understand the chemical combination and the properties of chemical substances more or less as we understand them now, freed from all the older implications of artificial gold or immortality elixirs. To his younger contemporary, J. B. van Helmont, I have already referred. We are now in the realm of the history of modern chemistry, which cannot be followed further here. Suffice it to name the great seventeenth-century chemists Béguin, Sala, Glauber, Becher, Stahl, Lémery and Boyle. Then the eighteenth century brings Boerhaave and Macquer, ending with Antoine Laurent Lavoisier—and 'la revolution en chimie est faite'.

xix

Introduction

The new movement introduced by Paracelsus got the name of iatro-chemistry because it was chemistry applied to medicine, and all biochemists therefore are particularly interested in this period. One has to understand the very mixed character of the Paracelsian contribution, for while on the one hand it called insistently for ever new experiments, and did indeed make many fresh discoveries in chemistry, it had at the same time a world-philosophy derived from Neo-Platonic, Gnostic and Hermetic sources. This was strangely similar to certain Chinese world-views involving the macrocosm-microcosm doctrine, the principle of action at a distance, the unification of the spiritual and corporeal worlds, the idea of the inter-connexion of all things by universal sympathies and antipathies, and a tendency to numerology as opposed to real mathematics and quantification. So far did this resemblance go that one of the English Paracelsians, Robert Fludd, invented two words, 'volunty' and 'nolunty', for which he could just as easily have chosen Yang and Yin if he had known of their existence. In time the Paracelsians or 'chymists' came into sharp conflict with the adherents of the traditional herbal pharmacopoeia, the 'Galenists', in a famous controversy towards the end of the seventeenth century.* In the Eastern civilizations this could not have happened because they had never had the prejudice against metallic and mineral remedies that Greek medicine had bequeathed to Latin Europe. What here concerns us more, however, is the fact that in China also there was an iatro-chemical movement, though it started considerably earlier. Since it led to what was perhaps the greatest achievement in all the pre-natal history of biochemistry, something must be said about it here.

In order to understand this, one must know that as Chinese

* These were the circumstances in which arose the first biochemical organization, so to say, in this country, the Society of Chymical Physitians, founded in 1665 by William Goddard, Marchamont Needham and others, just as the Great Plague was beginning to ravage London. It did not long survive that emergency, but the medical profession gradually·adopted by general consent those chemical remedies that experience showed to be effective. Since the subject does not arise elsewhere in this book, it may be of interest to add that in 1808 a group of Fellows of the Royal Society, including Humphrey Davy, Charles Hatchett, Everard Home and Benjamin Brodie, founded a Society for the Improvement of Animal Chemistry, which lasted a couple of decades. Finally in 1911 came the establishment of the present Biochemical Society, convoked by J. A. Gardner and R. H. A. Plimmer.

alchemy pursued its way during and after the T'ang period, it developed two entirely different strains of thought and activity, the 'outer elixir' (*wai tan*) and the 'inner elixir' (*nei tan*). The former was the standard term for all the longevity preparations made from metals and other inorganic substances as well as products of plants and animals. The latter signified the school of all those—and there were many, eventually preponderating in number—who said that nothing important would ever be effected by trying to act upon the human body pharmacologically from outside; what was necessary was to engage in a variety of special practices which would bring about the formation from the body's own tissues and juices of a truly organic elixir of immortality. These procedures were elaborate but essentially psycho-physiological; e.g. meditational, respiratory, gymnastic, photo-therapeutic, and sexual. They were all carried on in the conviction that 'the only true laboratory is the body of man himself'—that was to be the reaction-vessel in which the marriage of contraries, the *conjunctio oppositorum*, would take place. It will be seen that this Chinese system was quasi-yogistic, and it is easy to point to the Indian connexions, though the direction and extent of the transmissions still remains unclear. In any case the Chinese conceptions were distinctly more materialist and 'biochemical' than anything in the Indian mind, for the Chinese pictured the elixir formed in the body as a definitely chemical thing, a 'self-made medicine'. Although in the later centuries there were great exponents of practical, laboratory, inorganic alchemy such as Ts'ui Fang and Wang Chieh in the eleventh century, and P'êng Ssu around 1225, most of the great names belong to the *nei tan* tradition, e.g. Chang Po-Tuan (d. 1082) and Ch'en Chih-Hsü (*c.* 1330). From all this it can be seen that Chinese *nei tan* alchemy had nothing to do with the allegorical-mystical-psychological trend which tended to dominate in European alchemy, especially as its sands ran out and modern chemistry began to take over.

And now comes the dénouement, for in the Chinese iatro-chemical movement, current from the Sung to the beginning of the Ch'ing dynasties (eleventh to seventeenth centuries), the methods of the *wai tan* alchemists began to be applied, especially under medical influence, to what one might call *nei tan* materials. Take urine, for example. Textual evidence shows in great detail that during the centuries just named the Chinese iatrochemists worked up large quantities of it almost as we should say 'on a manufacturing scale'. And they subjected it to evaporations, heatings, precipitations (even with saponins), re-solution and further precipitations, ending with sublimation procedures at strictly controlled temperatures. In this way they were able to prepare, centuries before the nature of the steroid ring-system could have been conceived of, mixtures of crystalline steroid sex-hormones, and these the physicians used for just the same conditions as those in which they are prescribed today. By choosing the age and sex of the donors varying mixtures of the crystalline products could be obtained. The whole process must be called quasi-empirical rather than empirical because some fairly sophisticated theories were at the basis of it, though they were not those of modern science. This *ch'iu shih* is only one example of the Chinese iatro-chemical preparations, for many other materials were also worked up, e.g. placenta, menstrual blood, testis and thyroid gland; animal sources being often used. This was what I ventured to call perhaps the greatest achievement in the pre-natal history of biochemistry.

In order to complete the picture, something must be said of the Indian contribution. It seems on the whole to have been truly alchemical from the very beginning, combining the idea of gold-making with that of the elixir (*rasa*). On the other hand it does not seem, so far as we can tell at present, to be as old as Hellenistic proto-chemistry, let alone the alchemy of China. The beginnings are over-shadowed by the figure of the Buddhist patriarch Nāgārjuna, who may or may not have been the eminent alchemist that later generations thought he was; perhaps there was another person of the same name working in the third or fourth century A.D., the latter time being the date of the earliest document, the Bower MS. Very little early mediaeval literature has remained, however, and it is only after the *Rasaratnākara*, which is dated to the seventh or eighth century, that texts become at all numerous. The bulk of the literature is rather of the tenth to the fifteenth centuries. Elucidating the development of Indian alchemy as such is a particularly difficult task, partly because the dating of any document in Indian culture is a ticklish matter, and partly because a wealth of MSS still lie unexamined in the Indian libraries. Particularly interesting results are to be expected from the further study of the Tamil literature of the south. In general one can hardly say more for the present than that the contacts between the Indian, Chinese and Arabic culture-areas were often quite close, but that India remains still by far the most obscure of the three.

Only two more key-words now remain to be considered— Conjunction, and Ferment. The idea of the conjunction of opposites in synthetic reaction has already appeared in passing as the aim of *nei tan* alchemists, but in fact it was the objective of all alchemists everywhere, and nowhere was it more enthusiastically sought than in the Latin world. It is of course nothing less than the ancestor of the modern doctrines of chemical affinity, fundamental in biochemistry as elsewhere; yet its roots go right back to the ideas about sympathies and antipathies which preoccupied Liu An about 125 B.C. no less than his senior, Bolus of Mendes, and his junior, Pseudo-Democritus. Whether like tends to react with like, or like with unlike, was a problem that worried everybody for two millennia before Lavoisier's time; the Paracelsians and the Galenists tended to take opposite sides in the matter, and the Arabs had also discussed it. About the sixth century the Chinese took a notable step forward in sophistication by arranging chemical substances, including some of biochemical interest, in categories (*lei*) as well as the vertical Yin-Yang classification, members of the opposing sides of which must necessarily be expected to react. But if they did not belong to the same category (*t'ung lei*) then no reaction would take place. Such are some of the ancient ideas, originally always with a sexual nuance, from which men slowly passed to our more exact conceptions of valency, affinity, steric hindrance and the like; recognizable still, it may be, even in the activity constants of enzyme proteins (cf. lecture 2), and the points of action, whether stimulatory or inhibitory, of hormone molecules (cf. lecture 5).

It was not so many years ago that an enzyme was known as a

Introduction

Ferment, and this is the last key-word which there is time to consider here. In biology the idea of a small amount of leaven leavening a great mass was taken over simply from the empirical human technology of beer and bread; the 'domestication' of yeasts which must go back at least to Babylonian times. Very early the development of embryos was seen as a fermentation; morphological differentiation, with the appearance of complex organs, muscles, nerves and vessels, being analogized in a rather simple-minded way with the varied textures, shapes and colours which appear in maturing cheese. In the Jewish Wisdom Literature, produced just at the time when the biological school of Alexandria was at its height, Job (x. 10) is made to say 'Hast thou not poured me out as milk and curdled me like cheese? Thou hast clothed me with skin and flesh, and knit me together with bone and sinews'. Aristotle had said the same thing—for him the menstrual blood was the material basis of the foetus and the semen provided the form, acting upon it just as rennet acts upon milk. This idea, though followed little further, remained a commonplace throughout the Western Middle Ages, prominent for example in the visions of Hildegard of Bingen (1098 to 1180), and related to what Albertus Magnus had in mind when he said that 'eggs grow into embryos because their wetness is like the wetness of yeast'. This line of thought when further traced leads directly into the first work on the nature of proteins. The fascination which the Alexandrian proto-chemists had felt about the yolk and white of eggs a millennium and a half earlier, was felt again by Sir Thomas Browne (1605–82) in his chymical elaboratory at Norwich in the middle of the seventeenth century. For there he carried out many experiments with the chemical apparatus and equipment of the period to try and find out more about the proteinaceous substances which seemed endowed with such marvellous potentialities. Work paralleling this on the amniotic and allantoic fluids was done in 1667 by Walter Needham, and further efforts to unveil the secrets of the proteins of eggs were discussed at length by Hermann Boerhaave in his *Elementa Chemiae* of 1732, but no real break-through in the understanding of protein structure was possible of course until the development of organic chemistry in the nineteenth century, and the classical work of Emil Fischer stood nearer the end of that than the beginning.

What we have been talking about is nothing less than the beginning of man's knowledge of the phenomena of catalytic action; but this has a separate, rather different root, concerned not with biological events of any sort but with metallurgical techniques. In fact it goes back to what we should now call simply the 'debasement' of gold. The Graeco-Egyptian papyri of the third century constantly speak of the imitation of the precious metals by alloying them with others such as copper, tin, zinc and lead, but at least some of the gold and silver mixture (*asem, electrum*) had to be there, though quite a small amount would do. Hence the philosophers talked of the *diplosis* or *triplosis* (doubling or tripling) of *asem*, the necessary amount of which was sometimes thought of as a 'seed' or, as we might say, a nucleus of crystallization. However, in the Hellenistic proto-chemical corpus (first to seventh centuries) the idea that a small amount of something can act on a large amount of material to turn it into precious metal, the process that was later called projection, is already fully present, and the action of the small amount of substance (corresponding to the later 'philosopher's stone') tended rather to be likened to the action of yeast (*maza, azymon*). So for Zosimus (*c.* 300) the effect of projection was a fermentation. It is an extraordinary yet little-known fact that exactly the same idea of projection can be found as early as the first century B.C. in China; but whether we are to think of a Westward transmission or of some intermediate common source remains as yet quite uncertain. From the Chinese literature, and from the Hellenistic-Byzantine writings, the idea of the metallurgical 'ferment' passed into the alchemy of the Arabs, where nearly all the writers have it—the Jābirian corpus and Ibn 'Umail, also the *Turba* and the eleventh-century 'Book of Alums and Salts'. Thence to Geber about 1300 and to the Villanovan corpus of the early fourteenth century, where *massa* has become the regular word for the philosopher's stone as ferment, actual gold or silver still being thought of as necessary elements in its composition. Indeed a garbled etymology at times derived *alchemia* itself from *archymum* (*azymon*), so that all chemistry was synonymous with the 'yeasty craft' (*maza pragma*), as in the early fourteenth-century Byzantine Greek translation of Pseudo-Albertus. Now this replication process whereby a certain thing could make more of itself *ad infinitum* was surely a strange and not often recognized ancestral foreshadowing of the knowledge we now have about the self-replicating ribonucleo-proteins. Perhaps it well exemplifies the truth that 'the alchemists took a path just the opposite of chemistry today, for while we seek to explain biological processes in terms of chemical ones, they conversely explained inorganic phenomena in terms of biological events'.

There is still a little more to say about ferments, however, because the ancients and the early mediaeval people knew about digestion and the fundamental changes which it could carry out at strikingly low temperatures. All the common phenomena of fermentation and putrefaction were involved in their thinking about this. Hence even when they were in hot pursuit of the philosopher's stone, they were sometimes convinced that certain processes ought to be carried out slowly at low temperatures; which explains the frequent use among the mediaeval alchemists of fermenting horse-dung, by the aid of which temperatures of some 65 degrees could be maintained for prolonged periods. It was the same preoccupation which inspired the Chinese alchemists of the Sung to devise extraordinary systems of cooling coils in their reaction-vessels, so that they could control the temperatures attained. More fanciful was the tendency of some Western alchemists, especially the writers of the Lullian corpus of the late fourteenth century, to give their vessels the shapes of organs of the body—the stomach, the uterus, or the egg—and the age-old sexual nuance of chemical combination was mirrored in the paired alembics used at this period for reflux distillations. In the second iatro-chemical phase the nonsense was largely cast off, and van Helmont laid down that all the reactions of organic life were controlled by ferments of one kind or another. Disease also was thought by him and his followers to be due to 'alien ferments', their new version of the *contagium vivum*. Van Helmont recognized in the body (not only in the gut) six primary digestions, and here again the hidden parallelism between Chinese and Western thought reveals itself, for in 1624 van Helmont's contemporary, Chang Chieh-Pin, was explaining the doctrine then quite old in China, of the 'three coctive regions' (*san chiao*). Such correspondences invite much further research and I cannot say more of them here. But we have seen enough to realise that the idea of organic catalysts (cf. lectures 1, 2, 7) goes far back into the past, involving all the aspects of proto-chemistry and alchemy through the long course of the centuries.

This brings me to the end of what there is space to include by way of background panorama for the lectures which follow. The fact that they are now gathered together in print is very largely due to the persistence of Frank Young, the present head of the Sir William Dunn Institute of Biochemistry in Cambridge, who could not bear to see so interesting an assembly of perspectives restricted to oral tradition. I cannot end, however, without a filial tribute to that great man of whom most of us are the descendants, Frederick Gowland Hopkins, the *fundator et primus abbas* of biochemistry in this country. All of us who were privileged to follow his lectures for many years will remember how he never lost an opportunity of referring to events of historical interest in our subject. Whether it was van Helmont capturing the 'wild gas' in breweries, or Captain Lind protecting his men from scurvy with citrous fruits, or Berzelius finding lactic acid in the muscles of hunted stags in 1808, or Fick and Wislicenus romantically climbing the Faulhorn to settle the relation of protein metabolism to muscular work—all were brought in to give colour and perspective, to enlarge our understanding of our past. The present book may thus be considered one more laurel-wreath for 'Hoppy', yet at the same time adequate enough, we hope, to these present days of electron microscopy, refrigerated ultracentrifuges, the refinements of chromatography and electrophoresis—and the examination of Moon dust or Mars dust for forms of life that had not been dream'd of in our philosophy.

Introduction

REFERENCES

Berthelot, M. (1885). *Les origines de l'alchimie*. Steinheil Paris; repr. Libr. Sci. Arts, Paris (1938).

Berthelot, M. *Introduction à l'étude de la chimie des Anciens et du Moyen-Age*. First pub. as vol. 1 of Berthelot & Ruelle; repr. Libr. Sci. Arts, Paris (1938).

Berthelot, M., Duval, R. & Houdas, M. O. (1893). *La chimie au Moyen-Age*. 3 vols. Impr. Nat. Paris; repr. Zeller, Osnabrück, and Philo, Amsterdam (1967).

Berthelot, M. & Ruelle, C. E. (1888). *Collection des anciens alchimistes Grecs*. 3 vols. Steinheil, Paris; repr. Zeller, Osnabrück, 1967.

Coley, N. G. (1967). The Animal Chemistry Club; Assistant Society to the Royal Society. *Notes & Records Roy. Soc.*, **22**, 173.

Debus, A. G. (1964). The Paracelsian aerial nitre. *Isis*, **55**, 43.

Debus, A. G. (1964). Robert Fludd and the use of Gilbert's *De Magnete* in the weapon-salve controversy. *Journ. Hist. Med. All. Sci.* **19**, 389.

Debus, A. G. (1965). The significance of the history of early chemistry. *Journ. World Hist.* **9**, 39.

Debus, A. G. (1967). Fire analysis and the Elements in the 16th and 17th centuries. *Ann. Sci.* **23**, 127.

Debus, A. G. (1968). Mathematics and Nature in the chemical texts of the Renaissance. *Ambix*, **15**, 1.

Debus, A. G. (1968). *The chemical dream of the Renaissance*. Heffer, Cambridge.

Dubs, H. H. (1947). The beginnings of alchemy. *Isis*, **38**, 62.

Dubs, H. H. (1961). The origin of alchemy. *Ambix*, **9**, 23.

Eliade, M. (1954). *Le yoga, immortalité et liberté*. Payot, Paris.

Eliade, M. (1956). *Forgerons et alchimistes*. Flammarion, Paris. Eng. tr. Corrin, *The forge and the crucible*.

Filliozat, J. (1949). *La doctrine classique de la médecine indienne*. Imp. Nat., CNRS and Geuthner, Paris.

Ganzenmüller, W. (1938). *Die Alchemie im Mittelalter*. Paderborn. Repr. Olms, Hildesheim (1967).

Holmyard, E. J. (1957). *Alchemy*. Lane, London.

Jung, C. G. (1953). *Psychology and alchemy*. Routledge and Kegan Paul, London. (Coll. Wks, vol. 12.)

Leicester, H. M. (1965). *The historical background of chemistry*. Wiley, New York.

Levey, M. (1959). *Chemistry and chemical technology in ancient Mesopotamia*. Elsevier, Amsterdam and London.

Li Ch'iao-P'ing (1940). *Chung-Kuo Hua-Hsüeh Shih*. Ch'ang-sha. 2nd ed. rev. and enlarged, T'ai-pei (1955); Eng. tr. *The chemical arts of Old China,* Journ. Chem. Ed. Easton, Pa. (1948).

Lieben, F. (1935). *Geschichte d. physiologischen Chemie.* Deuticke, Leipzig and Vienna.

Lu, G.-D. & Needham, J. (1939). A contribution to the history of Chinese dietetics. *Isis,* pr. 1951, **42**, 13.

Lu, G.-D. & Needham, J. (1964). Mediaeval preparations of steroid hormones. *Med. Hist.* **8**, 101.

Lucas, A. (1948). *Ancient Egyptian materials and industries,* 3rd ed. Arnold, London.

Multhauf, R. P. (1954). John of Rupescissa and the origins of medical chemistry. *Isis,* **45**, 359.

Multhauf, R. P. (1956). The significance of distillation in Renaissance medical chemistry. *Bull. Inst. Hist. Med.* **30**, 329.

Multhauf, R. P. (1967). *The origins of chemistry.* Oldbourne, London.

Needham, J. (1934). *A history of embryology.* Cambridge; 2nd ed. revised, Abelard-Schuman, New York (1959).

Needham, J. (1954–) (with Wang Ling, Lu Gwei-Djen, Ho Ping-Yü, K. Robinson, Ts'ao T'ien-Ch'in *et al.*). *Science and civilisation in China.* 7 vols. in 11 or 12 parts, Cambridge.

Needham, J. (1962). Frederick Gowland Hopkins. *Notes and Records Roy. Soc.* **17**, 117; *Perspectives in Biol. and Med.,* **6**, 2.

Needham, J. (1967). The roles of Europe and China in the evolution of oecumenical science. *Adv. of Sci.* **24**, 83; *Journ. Asian Hist.* **1**, 3.

Needham, J. & Lu, G.-D. (1966). Proto-endocrinology in mediaeval China. *Jap. Studs. Hist. of Sci.* **5**, 150.

Needham, J. & Lu, G.-D. (1968). Sex hormones in the Middle Ages. *Endeavour,* **27**, 130.

Needham, J. & Pagel, W. (ed.) (1938). *Background to modern science; ten lectures at Cambridge arranged by the History of Science Committee* (by F. W. Aston, W. L. Bragg, F. M. Cornford, Wm. Dampier, Arthur S. Eddington, J. B. S. Haldane, G. H. F. Nuttall, R. C. Punhett, Rutherford and J. A. Ryle). Cambridge.

Pagel, W. (1944). The religious and philosophical aspects of van Helmont's science and medicine. *Bull. Inst. Hist. Med.* Suppl. no. 2.

Pagel, W. (1958). *Paracelsus; an introduction to philosophical medicine in the era of the Renaissance.* Karger, Basel and New York.

Pagel, W. (1962). The 'Wild Spirit' (gas) of J. B. van Helmont, and Paracelsus. *Ambix,* **10**, 2.

Pagel, W. (1968). Paracelsus; traditionalism and mediaeval sources, art. in *Medicine, science and culture,* p. 51. Temkin Presentation Volume, ed. L. G. Stevenson and R. P. Multhauf. Johns Hopkins Press, Baltimore, Md.

Introduction

Partington, J. R. (1961–). *A history of chemistry*, 4 vols. Macmillan, London.

Plimmer, R. H. A. (1949). *The history of the Biochemical Society*. Cambridge.

Ray, P. C. (1956). *A history of chemistry in ancient and mediaeval India*, ed. P. Ray. Ind. Chem. Soc. Calcutta.

Rex, F., Atterer, M., Deichgräber, K. & Rumpf, K. (1964). *Die Alchemie des Andreas Libavius, ein Lehrbuch der Chemie aus dem Jahre 1597, zum ersten mal in deutscher Übersetzung herasugegeben*. Verlag Chemie, Weinheim.

Ruska, J. & Kraus, P. (1930). Der Zusammenbruch der Dschābir-Legende; die bisherigen Versuche, das Dschābir-problem zu lösen—Dschābir ibn Hajjān u.d. Isma'īlijja. *Jahresber. d. Forschungsinst. f. Gesch. d. Naturwiss. (Berlin)*, **3** (9), 23.

Singer, C. (1931). *A short history of biology*. Oxford.

Sivin, N. (1968). Preliminary studies in Chinese alchemy; the *Tan Ching Yao Chüeh* attributed to Sun Ssu-Mo (*ca*. 581 to after 674), Inaug. Diss.; pub. as *Chinese alchemy; preliminary studies*. Harvard Univ. Press. Cambridge, Mass.

Stillman, J. M. (1924). *The story of alchemy and early chemistry*. Constable; repr. Dover, New York (1960).

Taylor, F. Sherwood (1930). A survey of Greek alchemy. *Journ. Hellenic Studs.* **50**, 109.

Taylor, F. Sherwood (1937). The origins of Greek alchemy. *Ambix*, **1**, 30.

Taylor, F. Sherwood (1945). The evolution of the still. *Ann. Sci.* **5**, 185.

Taylor, F. Sherwood (1951). *The alchemists*. Heinemann, London.

Taylor, F. Sherwood (1953). The idea of the quintessence, art. in *Science, medicine and history*, vol. 1, p. 247. Charles Singer Presentation Volume, ed. E. A. Underwood, Oxford

Taylor, F. Sherwood (1957). *A history of industrial chemistry*. Heinemann, London.

Temkin, O. (1955). Medicine and Graeco-Arabic alchemy. *Bull. Inst. Hist. Med.* **29**, 134.

Thomas, Sir H. (1953). The Society of Chymical Physitians; an echo of the great plague of London, 1665, art. in *Science, medicine and history*, vol. 1, p. 56. Charles Singer Presentation Volume, ed. E. A. Underwood, Oxford.

Wasson, R. G. (1968). *Soma, divine mushroom of immortality*. Harcourt Brace and World, New York; Mouton, the Hague.

Webster, C. (1967). English medical reformers of the puritan revolution; a background to the Society of Chymical Physitians. *Ambix*, **14**, 16.

1

THE GROWTH OF OUR KNOWLEDGE OF PHOTOSYNTHESIS

by Robert Hill

There is not as much green shade now as in former days. Before I attempt to take you back with the green thoughts let us look at these two diagrams of a great country (figs. 1, 2). There is an interval of ninety-five years. We use enormous quantities of wood-pulp now. All this is only one reason why the study of photosynthesis is important.

Photosynthesis is the word used to describe a process fundamental to plant nutrition. This word came into being relatively recently. Absent in the 1880 edition of Pfeffer's great *Plant physiology*, it appeared in the 1897 edition as 'photosynthetische assimilation'. It was translated by Ewart as 'photosynthetic assimilation' (Pfeffer, 1900–1906). Between these two editions, Barnes in America (1893) had proposed the word 'photosyntax' and had rejected 'photosynthesis' on etymological grounds; one of his colleagues, Macdougal, preferred photosynthesis and he used this form in his textbook on practical plant physiology. The true origin of the word does not seem to be stated in any of the numerous textbooks that we have. Judging from the literature there seems to have been a polemic about who had actually invented it.

At that time, during the second half of the nineteenth century, the process of photosynthesis was formulated as:

Carbon dioxide + water + light =
Carbohydrate + oxygen + chemical energy.

This was essentially the reverse of the process of respiration. The green plants are called autotrophic or self-feeding because they require to take in no organic food for the increase of their substance. Because animals, including human beings, are

Robert Hill

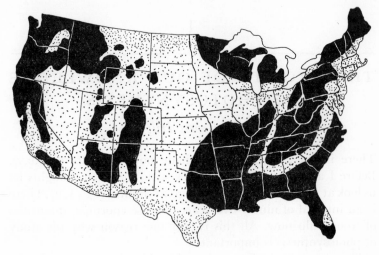

Fig. 1. Virgin forests in 1850

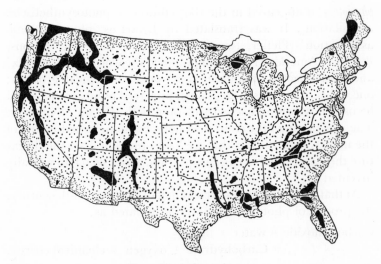

Fig. 2. Virgin forests in 1945

(Both figs 1 and 2 from R. A. H. Thompson, 'What's happening to the timber'
Harper's Magazine, August 1945, p. 125. Reproduced in H. Gaffron 'Photosynthesis
and the production of organic matter on Earth' in *Currents in biochemical research*,
ed. D. E. Green, New York, Interscience, 1946)

2

absolutely dependent on their coexistence with plants our thoughts go back to the very dim past. E. J. H. Corner (1964) once stated that the subject of botany is essentially the relation between man and plants—botany is a very old subject indeed.

In prehistoric times, in places where agriculture was developed, it would be clear that water and soil were both essential requirements for the living plants. It would have also become obvious that plants could be burnt, for, when the use of fire had been established, wood was used as fuel.

In considering the study of plant nutrition, I can attempt to take you back to the beginning of the Greek era. The essay in this is inevitably distorted because, as you no doubt will allow, we are so conditioned with scientific jargon and later conceptions. But you could see that the Greeks would grasp a relation between the plant and one at least of the items in our equation. That is the *water*. Nowadays, we could feel that there was some dim connexion between a prehistoric conception of fire and perhaps what we now call energy. But energy in the precise scientific sense is a very recent conception; Partington gives 1850 as the date, virtually the last item to be placed correctly in the nineteenth-century formulation.

At any time when people simply think about things, what is variously held to be right and true shows wide discrepancies. So it was that quite independent and mutually conflicting views about the nature of things have come down to us from the Greeks. These influences, sometimes quite amazingly, are still to be recognized in current thought. In science we assume that the thought is tempered by results of experiments and by the experimental methods. It would be fine to have a thought that would explain everything.

Thales of Miletus (*c.* 640–546 B.C.) gave us the thought 'all things are water' (πάντα ὕδωρ ἐστί)

Water = Plant.

It seems now that when this thought took over there would have been a subjective sense of ultimate reality. Could it not

3

somehow resemble in effect the equation of Einstein relating energy and mass? Perhaps in this way we could follow Parmenides of Elea (b. *c.* 539 B.C.) who considered the 'one' (which *is*) and the 'not one' or the 'many' (which *is not*) as being separate for investigation. He was considered by Plato (427–347 B.C.) to be not an idealist but a precursor for idealism. But Plato had maintained that to study separately the 'one' and the 'many' was sterile. An approximation to this as we might see it now, being precocupied with science, would be: the 'one' is a joy and a stimulus, the 'many' is a toiling and a moiling.

A different type of theory which was associated with Democritus, *c.* 360 B.C., was that everything was composed of characteristic atoms. These atoms would confer the characteristic to each plant. Here the atoms composing an olive tree would generally be supposed to be different from those composing the grape vine, that is Atoms (*a*) = Plant (*a*), and Atoms (*b*) = Plant (*b*). This was an important idea in relation to a Greek conception of plant mutations which persisted through the middle ages. The plants took their nourishment from the earth. Each plant took the atoms appropriate to its kind. Thus the plants, unlike animals, produced no excreta. But for animals the conception was that the food, after intake, was absorbed in part by an intelligent agent and the inappropriate parts were rejected. For the plant, by contrast, the intelligent agent would have to be in the earth or soil. Thus a plant would have seemed to resemble a foetus drawing its nourishment from the Mother-Earth, where the appropriate atoms were elaborated in the soil. We might note how the botanist F. O. Bower (1855–1948) in his book *Botany of the living plant* described a growing plant as showing continuous embryology. Again, do not some of the still current non-scientific ideas about humus and plant growth seem to be derived from this Greek idea of plant nutrition?

Earth = Plant.

The other conception of the plant being derived from water also persisted. This involved a transmutation theory even though most living plants like animals consist of 50–90 per cent. of water. These two representations of nutrition of plants had a significant influence extending in to the era of direct scientific experiment in the sixteenth to eigthteenth centuries. In fact they tended to be reconciled by a widely held belief that water would change into earth. Water = Earth.

Bringing us back to Thales of Miletus again.

Then there is the theory of the four elements (see fig. 3) which seems to have been brought in to greatest development

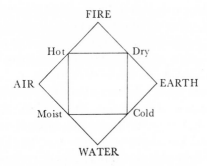

Fig. 3. Aristotle's theory of the four elements

(From J. R. Partington (1957) *A short history of chemistry*, 3rd ed. London: Macmillan)

by Aristotle. This is sometimes considered to have been a somewhat negative influence on scientific development, on account of the fire principle being considered a material substance. Be that as it may, the influence was powerful—it happens that just as of old, some of us are more interested in people than in things. The magnificent effort towards perfection in all things was especially accentuated in Plato who was interested in the individual and the society. Concepts of reality in this way were relevant to men's relations with each other but less useful in men's relation with the earth or 'Nature'.

The four elements were defined by Aristotle in much the same way as originally were the chemical elements of the present day. The differences in what we now understand as a

5

mixture from that of a single compound was appreciated. Mercury metal was called silver-water; the different kinds of matter were described in terms of intermediate or combined qualities. It is in this question of matter and quality of attributes where the difficulties for scientific progress came in. Nowadays we should tend to think that the four elements would symbolize a manifestation of energy as fire, and the three states of matter, solid, liquid and gas. In any case the idea of transmutation became dominant in the alchemical middle ages. In evaporation something changed into air and, in liquifacture of a solid, part of the earth had to be changed into water. With the original atomic ideas there was no need, I think, for the atoms to change into each other. The conservation of material seems to have been taken as axiomatic. The changes of state were readily appreciated by Heron of Alexandria in the first century A.D. In fact from what is known about his writing, as also from the *De rerum natura* of Lucretius, the modern theory of gases was anticipated. Yet, it was many hundreds of years later that an experimental knowledge of gases was developed.

Quite apart from this theoretical aspect very much practical experience on both agriculture and in chemical matters existed in early times. This can be gathered from the writings of Pliny the elder especially as regards soils and fertilizers. Again it was long known that it was often dangerous to go in to the bottoms of wine vats. If, on testing with a lamp, the lamp did not go out, this showed that it was safe. A famous experiment of a candle burning in a closed space over water was carried out by Philon of Byzantium (third century B.C.). After the candle became extinguished a contraction was observed. But experiments and material practice were usually subsequent to theory and often diverged from it.

The growth of our knowledge of photosynthesis

THE BEGINNINGS OF MODERN EXPERIMENT

Just as the Greek philosophers had sought for perfection in the individual and society so the early alchemists sought for perfection in material. Now we often think of the pursuit of alchemy as being the effort to transmute dross into gold. On the other hand the search for the philosopher's stone can be viewed as an effort to find an imagined 'one way' to control material processes. The gap which had so long appeared to exist between practical men and the philosophers gradually disappeared. The study of natural philosophy came to be directed by the results of experiments, rather than by systematic description and cataloguing. In *The legacy of Greece* D'Arcy Thompson (1921) in his account of Aristotle discussed the reasons for the absence of contact between trade and scholasticism. Even up to the time of Robert Boyle (1627–91) chemistry does not seem to have been regarded as a 'proper' study, being too closely associated with a menial task.

The experimental method for the study of plant nutrition is generally considered to begin with van Helmont. He not only invented the word gas, *c.* 1630, but also contributed much to the beginnings of chemistry, as distinct from alchemy, and again, indeed, in view of his conception of a 'ferment', to the beginnings of biochemistry. Van Helmont decided to test the theory:

$$\text{Water} = \text{Plant}.$$

His famous experiment, which was so carefully carried out, nearly completely satisfied the theory when judged by the knowledge at the time. The permanent value of the experiment lies in the fact that it showed how little of the actual matter composing the plant can be derived from the soil.*

I took an Earthen Vessel, in which I put 200 pounds of Earth that had been dried in a Furnace, which I moystened with Rain-water, and I implanted therein the Trunk or Stem of a Willow Tree, weighing five pounds; and at length, five years being finished, the Tree sprung from thence, did weigh 169 pounds, and about three

* The quotation is taken from Partington (1957), *A short history of chemistry*, pp. 51–2.

ounces: But I moystened the Earthen Vessel with Rain-water, or distilled water (alwayes when there was need) and it was large and implanted into the Earth, and least the dust that flew about should be co-mingled with the Earth, I covered the lip or mouth of the Vessel, with an Iron-Plate covered with Tin, and easily passable with many holes. I computed not the weight of the leaves that fell off in the four Autumnes. At length, I again dried the Earth of the Vessel, and there were found the same 200 pounds, wanting about two ounces. Therefore 164 pounds of Wood, Barks, and Roots, arose out of water onely.

Now the contemporary interpretation of this experiment, I think, could really involve the conflicting notions of conservation of the soil (earth) element and the transformation of the water element. Both the conservation and transmutation ideas had come from the Greek philosophy. Then again there was the idea of each plant requiring to take its specific kind of atoms from the soil. The different kinds of plants, especially those used in agriculture and in medicine, have been recognized, depicted and described from very early times. In fact, the purely descriptive side of botany has dominated in the study of plants on more than one occasion in the history of this science.

It was simply this description of plant species and not any direct experiment that was used by Edme Mariotte (*c.* 1620–84) to test a Greek view of the earth nutrition of plants. He argued thus. I can take a definite amount of soil say 7–8 lb. In this soil I can grow a plant which will weigh say ¼ lb.—but it matters little out of 3,000–4,000 different kinds of plant which one I choose to grow. Thus if each plant has to take atoms or elements for its characteristic kind from the soil (and a small amount of rain water with its salts) there certainly could not be enough kinds of atoms to produce the material of 3,000 different plants.

THE DISCOVERY OF PHOTOSYNTHESIS

As was indicated previously, from the time of Robert Boyle (1627–91) and the foundation of the Royal Society in England (1645–62) the pursuit of experimental science came to be considered quite respectable. Eugene Rabinowitch in volume 1 of his classical book on photosynthesis (1945–58) has given a splendid historical account of the discovery of photosynthesis starting from the work of Stephen Hales (1677–1761) up to the time of the famous plant physiologist Julius von Sachs (1832–97). Rabinowitch gives to Stephen Hales the credit for showing that air must be an important constituent of all organic matter. He quotes from *Vegetable staticks* (1727) where Hales wrote 'Plants very probably draw through their leaves some part of their nourishment from the air, may not light also by freely entering surfaces of leaves and flowers contribute much to ennobling the principles of vegetables'.

This is especially interesting because in previous descriptions of plants the requirement for light, apart from its affect on ripening fruit, seems hardly ever to have been mentioned. The necessity for light must have been obvious in very early days but perhaps only apparent to people working in horticultural practice.

It now seems that we should consider the work of van Helmont as providing much of the experimental basis for the development towards modern chemistry and biochemistry. He was not only able to distinguish the properties of different kinds of gas but also distinguished them by different methods of producing them. He burnt charcoal and called the product 'gas carbonum', which also he called 'gas silvestre' in circumstances where it comes from fermenting wine. This was the 'fixed air' later described by Joseph Black (1728–99) which he obtained from the action of acid on limestone. The word air continued to be used to designate the different gases for a considerable time after van Helmont.

It was never possible to contain a gas in an open vessel and in some experiments a closed vessel would burst. The development of the study of the nutrition of plants, and indeed the development of chemistry itself largely depended on the experimental procedures which involved gases.

The postulate of conservation of material substance seems to have been the basis of both the atomic and transmutational theories in antiquity. This perhaps accounted for the tendency to regard all the four elements as substances. It could long have been observed that flame was burning smoke; this was stated by Aristotle and by van Helmont. Thus fire came to be represented by a flame ($\phi\lambda o\gamma\iota\sigma\tau o\nu$) principle which had to be an actual material component of all combustible materials. We may perhaps see how this came about because heat itself was generally considered to be a fluid contained in materials. Count Rumford's experiment (1798) on the heat produced in boring cannons revived and proved the alternative theory, very long accepted by a minority, that heat was motion. The word 'caloric' dating from 1787 in France and the German 'warm-stoff' may still be found in (not so old) dictionaries. The postulation of a material flame principle led to the 'phlogiston' theory, which is associated with J. J. Becher, 1635–1682 and with G. E. Stahl, 1660–1734. On this theory, when the candle, being very rich in phlogiston, was burnt in a closed space the air became 'phlogistigated' and then it was incapable of supporting combustion. This strange scientific jargon actually persisted among a few scientists until the beginning of the nineteenth century.

The real beginning of the experimental study of photosynthesis was due to Joseph Priestley (1733–1804). He was skilful in carrying out experiments with gases—air as they were called even then in spite of van Helmont (see fig. 4). The great discovery was when Priestley found that the 'phlogistigated' air, which no longer could support combustion and the respiration of a mouse, could be restored by a green plant. The green plant could 'dephlogistigate' the 'phlogistigated air'. The gas which Priestley called 'dephlogistigated air' we should now, following Lavoisier, call oxygen. In this discovery Priestley first showed the complementary relationship existing between vegetation and animals. Priestley was especially amazed by the vigorous combustion of a candle in his dephlogistigated air when he prepared it from oxide of mercury. He realized that the common air was less pure. Partington (1957) quotes him: 'the air which Nature has provided for us is as good as we deserve'. Later, Priestley showed, by using his 'nitrous air' (nitric oxide), how to determine the amount of pure dephlogistigated air in the atmosphere. He found one-fifth of a volume present; this was an accurate result.

The growth of our knowledge of photosynthesis

The discovery by Priestley seemed to have initiated at least three independent workers in the subject. The first was Jan Ingenhousz (1730–99) who while acknowledging Priestley continued to follow up his work with success. Rabinowitch considers that Ingenhousz established that light was necessary

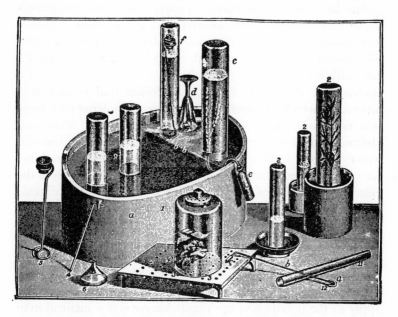

Fig. 4. Priestley's pneumatic trough and other apparatus

(From his *Experiments and observations on air*, Vol. 1, 1774. Reproduced in J. R. Partington (1957) *A short history of chemistry*, 3rd ed. London: Macmillan)

and that he had shown that 'something phlogistigated' was produced in the plant. Also it appeared that Jean Senebier (1742–1809) had discovered the requirement for fixed air, which originally had been denied by Ingenhousz. Later Nicolas Theodore de Saussure (1767–1845) obtained the first evidence that water was involved. We can be amazed at the progress made during this period of phlogiston and with the lack of a precise scientific terminology. In his *History of botany* (1890) Julius von Sachs (1832–97) gives the credit for the discovery of photosynthesis to Ingenhousz. This is no doubt connected with the fact that in 1796 a book entitled *Food of plants and renovation of the soil* was published where Ingenhousz had adopted the new chemical theory of Lavoisier. Dephlogistigated air was called oxygen and fixed air was called carbonic acid.

Antoine Laurent Lavoisier (1743–94) had carried out quantitative experiments which showed that the phlogiston theory was untenable. When a substance was burnt in air it combined with the oxgyen; so that a metal, when heated in air, formed a calx with increase in weight while the volume of air showed a contraction. Previously in 1773 Lavoisier had published a memoir in which he described experiments showing that there was no possibility of water changing into earth. Since van Helmont and his experiment it had been found that distilled water on evaporation always leaves a slight earthy residue. Lavoisier showed that the residue was derived entirely from the vessel by the action of the water. Thus it was the carbon dioxide in the air which had supplied the main nourishment to the five year old willow tree, showing how near van Helmont had been to the discovery of photosynthesis.

For the final stage in defining the process of photosynthesis the credit is given to Julius Robert Mayer (1814–78) for introducing the conception of conservation of energy as it would apply to the nutrition of the plant. Mayer was a medical man and his publication concerning the general application of energy conservation was considered by some other scientific men to have been premature. Walther Nernst (1864–1941) has given him credit in this remarkable statement (1916), 'On the ground which was richly fertilised by the ruined fancies of unlucky inventors, grew up like a tree of knowledge the law of the indestructibility of energy, the golden fruit of which was plucked by Mayer and Helmholtz'. The early experiments with plant nutrition were carried out at a time when the theory of nature of chemical changes was in its infancy. By the time J. von Sachs was writing his *Plant physiology* many of the processes involving chemical and physical changes could be described by precise scientific terms. He found that the first carbohydrate to be visibly formed in light was in many cases starch. Thus, from a general physiological viewpoint, the relation of photosynthesis to plant nutrition was accepted and it formed the basis for a variety of experimental works, with plants. However there seemed to be little progress as regards the actual biochemical mechanism of photosynthesis until the beginning of the twentieth century.

To begin with, the emphasis was all on the green leaf pigment. Chlorophyll was originally the name given by Pelletier and Caventou in 1818; this name included all the pigments found in the green leaf. When Tswett, at the beginning of this century, discovered how to separate the pigments by chromatography he used the term chlorophyllin for the two green pigments in the leaves which originally had been shown to be present by Sorby in 1873. However Willstätter and Stoll in their work on the leaf pigments and on photosynthesis published in 1913 and 1918 used the word chlorophyll for the unmodified leaf pigments. They used chlorophyllin to describe modified chlorophyll; this may have been due to the fact that in the calcium carbonate used for chromatography the green pigments were apt to suffer a chemical alteration. It had been widely held that chlorophyll was the agent in the plant wholly responsible for photosynthesis. Fruitless experiments were directed towards proving this while chlorophyll continued to be a main focus of attention. However it happened that in more recent times between 1930 and 1940 the focus of attention came back again to water. C. B. van Niel (1949) in the 1930s had carried out the most searching studies with photosynthetic bacteria. These do not produce oxygen, which distinguishes them from the green plants. The bacteria were found to cause oxidation of a substance in their environment, in the absence of atmospheric oxygen. Van Niel could express symbolically all the types of photosynthesis in one way. The result of the light absorbed by the pigments in the living cells was to resolve water, represented by HOH, into the parts H and OH. This could give rise to a reduction and a simultaneous oxidation. As a result, the active processes involving the reduction of carbon dioxide and oxygen production could be carried out by means of ferments (enzymes) in the dark. This theory after about 2,400 years (since Thales of

Robert Hill

Miletus) seemed to be a single thought which has motivated experiments extending even to the present time, leading towards a more complete understanding of the mechanism of photosynthesis.

REFERENCES

Bower, F. O. (1961). *Botany of the living plant*, 4th ed. New York: Hafner.

Corner, E. J. H. (1964). *The life of plants*. London: Weidenfeld and Nicholson.

Nernst, W. (1916). *Theoretical chemistry*, 4th English ed. Tr. from the 7th German ed. by H. T. Tizard. London: Macmillan.

Partington, J. R. (1957). *A short history of chemistry*, 3rd ed. London: Macmillan.

Pfeffer, W. (1900–6). *The physiology of plants*, 3 vols. 2nd ed. tr. A. J. Ewart. Oxford: Clarendon Press.

Rabinowitch, E. I. (1945–58). *Photosynthesis and related processes*, Vols. I, II 1 and II 2. New York: Interscience; see also F. Czapek, *Biochemie de Pflanzen*, 2nd edn. Vol. 1. Jena: Gustav Fischer (1913).

Thompson, D'Arcy W. (1921). In *The legacy of Greece*, ed. R. W. Livingstone. Oxford: Clarendon Press.

van Niel, C. B. (1949). 'The comparative biochemistry of photosynthesis' in *Photosynthesis in plants*, ed. J. Franck and W. E. Loomis. Iowa: Iowa State College Press.

von Sachs, J. (1890). *History of botany, 1530–1860*, rev. ed. by I. B. Balfour, tr. by H. E. F. Garnsey. New York: Russell and Russell.

2

THE HISTORY OF ENZYMES AND OF BIOLOGICAL OXIDATIONS

by Malcolm Dixon

I have to deal here with two subjects, but they are closely related, for biological oxidations are catalysed by enzymes. It is these processes within living cells that provide the supply of energy which keeps them going.

The living organism—the living cell—has sometimes been compared with a flame, and this comparison is instructive, for they have much in common. By way of introduction let us consider this briefly.

Both consist of a region in which a complex series of successive chemical reactions is taking place; in both, heat is produced, largely by oxidation reactions; both have a shape or form which persists although the matter which composes them is constantly changing, so that there is a continuous flow of matter through them, and the atoms of which they are composed now are not those of which they will be composed some time later; both require for their maintenance in a steady state a constant expenditure of energy, and they disintegrate when this is cut off by starvation or by turning off the gas.

There are, however, two fundamental differences. The first is that in the flame the reactions take place because the temperature is very high; in living matter the temperature is low and the reactions occur only because of the presence of a number of highly specific *catalysts*—the enzymes, whose history we are about to consider. These catalysts are numerous, running into many hundreds. They are also highly specific, so that as a first approximation we may say that each enzyme catalyses its own particular chemical reaction. It is a consequence of this specificity that you have in living matter ordered sequences of reactions, or metabolic pathways, which are strictly defined and determined by the particular enzymes that are present. In the flame you have chaos; in the cell you have order.

The second fundamental difference is that in the flame the reactions are breakdown reactions exclusively; whereas in living matter the pattern of the reaction sequences is such that it includes many synthetic reactions. The energy required for these is supplied by the oxidative breakdown reactions going on at the same time, and the transfer of energy from one set of reactions to the other is brought about through the network of enzyme-catalysed reactions. In this way the system of intracellular enzymes brings about the synthesis of the organic constituents of living matter, necessary for maintenance and growth, and this includes synthesis of the enzyme catalysts themselves.

Most living cells, like the flame, carry out their oxidative reactions by means of molecular oxygen, but the utilization of O_2 by cells ('cell respiration') is due to the catalytic action of a system of oxidizing enzymes.

After this brief introduction, let us consider how knowledge of these systems has developed.

HISTORY OF ENZYMES

Enzymology is a science which has in the main developed late. Although just a few of the fundamental observations were made long ago, towards the beginning of the nineteenth century, the great developments have come during the last two or three decades. Most of the history of the subject is therefore comparatively recent. In 1920 about a dozen enzymes were known. None had been isolated, and their nature was anything but clear. Now about 900 different enzymes are known, of which a great many have been at least partially purified and approaching 200 have been obtained pure and crystalline. The complete chemical structure of a few has been determined. In fact enzymology has now become one of the most active growing-points of biochemistry. It is nevertheless most instructive to look back to the beginnings of the subject.

The history of enzymes

Discovery

Certain manifestations of enzyme action had been known for a very long time, although it was not known what caused them, still less that they were brought about by enzymes, for example fermentation, digestion and respiration. The first discovery of an enzyme appears to be due to Payen and Persoz in Paris.* Payen was director of a sugar factory in Paris; it was he who discovered cellulose and also the decolorizing action of charcoal. Persoz, one of the discoverers of dextrin, was afterwards a Professor at Strasbourg and later still in Paris. Both were pupils of Vauquelin.

Payen and Persoz (1833) published a paper which gives a surprisingly modern impression. From an aqueous extract of malt they obtained by precipitation with alcohol a fraction which contained *something* which could convert starch into sugar. They showed that this something (which we should now call amylase) was inactivated by heating, which is one of the characteristic properties of enzymes. They did not know at that date exactly what the chemistry of the process was, but they were impressed by the *separation* of the soluble sugar from the insoluble covering of the starch grain, and they called their factor 'diastase', from the Greek word for separation, διάστασις. That name was kept for amylases for some time, but after a time it became used for enzymes generally, especially in France. In 1898 Duclaux proposed the use of the last three letters as a suffix '-ase' to form the names of enzymes, a practice which is now universal.

The next enzyme to be found was an animal one—pepsin. In the following year (1834) it was found that gastric juice could digest food in the test-tube, and in 1836 Schwann in Berlin extracted the active principle with acid from the stomach wall. He showed that it could be precipitated and called it pepsin.

It is interesting to note that these discoveries of enzymes antedated the first suggestion of the idea of catalysis, put

---

* Pictures of a number of the workers mentioned in this account will be found in plates 1–18.

Malcolm Dixon

forward by Berzelius in Sweden in 1837. He called catalysis 'a force which is different from the forces previously known to us'. The idea is put forward in a new section which he inserted into the third edition of his great textbook of chemistry. On account of its importance, it is worth quoting at some length (Berzelius, 1837). After mentioning a number of examples of catalytic processes, such as the fermentation of sugar by yeast, and the decomposition of hydrogen peroxide by animal fibrin or coagulated plant proteins (due to the presence of catalase, no doubt), he continues as follows (translated from the German, itself translated from the original Swedish):

It has thus been shown that many bodies, both simple and complex, both in the solid or dissolved state, possess the property of exerting an influence on complex bodies which is quite different from ordinary chemical affinity, causing thereby a rearrangement of the constituents of the body into other relationships, without necessarily taking any part therein with their own constituents, although that may happen occasionally.

This is a new force, belonging both to inorganic and to organic Nature, for evoking chemical activity; a force which is probably more widespread than one had thought up to now, and the nature of which is still concealed from us...Availing myself of a well known derivation in chemistry, I will call it the 'catalytic power' of bodies, and decomposition by its means I will call 'catalysis', just as we understand by the word analysis the separation of the constituents of bodies in virtue of ordinary chemical affinity. Catalytic power seems really to consist in this, that bodies can, by their mere presence, and not by their affinity, arouse affinities slumbering at this temperature, and as a result the elements in a complex body become rearranged.

He asks, but does not answer, two questions about specificity: 'whether different catalytic bodies acting on the same complex body produce different products' and 'whether bodies with catalytic power exert this on a large number of complex bodies, or whether, as seems likely at present, they catalyse certain bodies without acting on others'.

He concludes with remarkable foresight:

When we turn with this idea to living Nature, an entirely new light dawns for us. It gives us good cause to suppose that in living plants and animals thousands of catalytic processes are taking place between the tissues and the fluids, producing the multitude of dissimilar chemical compounds for whose formation from the common raw material, sap or blood, we had not been able to think of any cause, but which in the future we shall probably find in the catalytic power of the organic tissue of which the organs of the living body consist.

A whole century was to elapse, however, before this remarkably accurate prophecy was confirmed.

It is important to realize that it was very largely the action of enzymes that gave rise to the idea of catalysis, not the converse as is often assumed.

At that time it was not known that fermentation was due to living organisms. This was shown in the following year (1838) by Cagniard de Latour, and more thoroughly twenty years later by Pasteur. It was natural then that *all* such catalytic actions should be attributed to living micro-organisms. The word 'ferment' was applied indiscriminately to denote all such agents, with no distinction between living organisms like yeast and what we now call enzymes. Fermentum, the Latin word for yeast, was a contraction of fervimentum, from ferveo, to boil up, alluding to the boiling appearance produced by the liberation of carbon dioxide. The application of the word ferment to both cells and enzymes led to a good deal of confusion for a number of years, in fact from about 1860 to about 1895.

Pasteur regarded the chemical reactions occurring in fermentation as an essential part of the life processes of the yeast. This did not appeal to the chemists, who did not like the intrusion of vital forces into chemistry. As Traube put it, 'Chemistry may well explain physiological processes, but physiology cannot explain chemical processes'. They tended to adopt the Berzelius position, that the chemical reactions were due to catalysis by definite chemical substances.

Now Berzelius had simply *described* the process of catalysis, but said the cause was unknown. Two years later (1839) a theory of enzyme action in fermentation was put forward by Liebig, who is often considered to be the father of biochemistry, and who was later Pasteur's chief opponent. Pasteur regarded fermentation as part of the life processes; Liebig regarded it as associated with death and decay. His theory involved the idea of a mechanical transfer of the movements of a decomposing body (the ferment) to a resting body (the substrate). He says,

The atoms of a putrefying body, the ferment, are in a ceaseless movement; they change their position in forming new combinations. These moving atoms are in contact with the atoms of sugar, whose elements are held together as sugar by a weak force. The movement of the atoms of the ferment cannot be without effect on the atoms of the elements of the substance mixed with it; either their motion is abolished or the atoms of the latter move too. The sugar atoms suffer a displacement; they arrange themselves in such a way that they hang together more firmly, so that they no longer follow the impact, that is to form alcohol and carbon dioxide.

Thus the sugar was almost literally shaken to pieces.

It is interesting that throughout the early history of the subject there seems to have been no suggestion that catalysis might go on *inside* the living cell. Berzelius spoke of a surface catalysis exerted by the tissues on the surrounding fluids, or by 'an insoluble body which we call "ferment"'. Liebig and others clearly regardèd fermentation as occurring at the surface of the yeast cells. The digestive enzymes too, which began to be discovered in that period, acted outside cells. Even the oxidative processes of respiration were thought to be extracellular, and in fact were believed to occur not in the tissues but in the blood of the lungs. Mayow had said in 1674 that 'With respect then to the process of respiration, it may be affirmed that an aerial *something*', (meaning oxygen) 'whatever it may be, essential to life, passes into the mass of the blood...Animals and fire draw particles of the same kind from the air...The heat of the blood arises I think from the effervescence of nitro-aerial particles with salino-sulphureous particles of the blood', which being translated into modern terms means 'the oxidation of combustible molecules of the blood by molecular oxygen'.

Of the small number of enzymes known in about 1850, I believe none was intracellular, and all enzyme reactions were apparently thought of as occurring outside living matter.

The history of enzymes

Knowledge of intracellular enzymes probably began with Berthelot (1860), who macerated yeast and was able by extraction and precipitation with alcohol to obtain an enzyme which inverted, or as we should say hydrolysed, cane sugar. He therefore put forward the quite definite view that this was one of a number of ferments *contained* in yeast, that it resembled the diastase of Payen and Persoz and the digestive ferments, that the living being itself is not the ferment but the producer of it, and that soluble ferments, once produced, function independently of any vital activity (an important point).

A great deal of the work on enzymes for the next sixty years was done on this enzyme under various names: the glucosic ferment, invertase, invertin, saccharase, sucrase, β-h-fructosidase, β-fructofuranosidase. It was long before the serious study of intracellular enzymes in general began, but it is the work on these enzymes from 1920–30 onwards that has brought about the vast increase in the number of known enzymes and the great increase in the understanding of living matter.

Meanwhile, in the mid-nineteenth century, it became fairly obvious that the digestive ferments were not micro-organisms, and that therefore ferments were of two kinds. These were distinguished by the names 'organized ferments', such as yeast and micro-organisms, and 'unorganized ferments', which we should now call enzymes. This nomenclature was rather unsatisfactory, and in 1878 Kühne introduced the name 'enzyme', from ἐν ζύμῃ (in yeast, with the stress on the 'in'), because these things were inside the yeast and not the yeast itself. He emphasized that the name did not imply that they were not present in more complex organisms too. This name is now universally adopted, although the name 'ferment' has lingered on in some countries almost to the present day. Incidentally, ζύμη comes from ζέω, to boil up, so that the derivation of the name corresponds exactly with that of the name ferment in Latin.

Finally Eduard Buchner in 1897 was able to extract from yeast a cell-free juice containing the whole fermentation system, showing that, contrary to Pasteur, living cells are not essential for fermentation.

Nature

The nature of enzymes remained for long obscure. In 1877 Traube put forward a general theory of ferment action, to include fermentation, cell oxidations, putrefaction etc., namely that ferments are substances allied to proteins which, while they themselves remained unchanged, were responsible for all the vital chemical changes in both higher and lower organisms. It is interesting that this correct view was suggested so early in the history of the subject, but there was little evidence of the protein nature of enzymes, and Traube's suggestion was not accepted for another half-century. No further real development was possible until a number of enzymes had been obtained in the pure state.

Meanwhile a good deal of doubt developed as to their protein nature, or indeed as to their being substances at all. Many workers, especially de Jager (1890) and Arthus (1896), held that enzymes are not chemical individuals, but that various kinds of bodies may have conferred upon them properties which cause them to behave like enzymes, so that we have to deal with properties rather than with substances. I can myself recall that when I began to receive instruction in the subject, enzymes were regarded more as a kind of influence or force which still hung about matter which had once been living, and even in the 1929 edition of the Encyclopaedia Britannica we read 'enzymes were formerly thought to be proteins, but this is no longer believed'. Barendrecht (1904) believed that they were radioactive bodies, the chemical activities being due to the radiations they emitted.

About the end of the nineteenth century Bertrand maintained that each enzyme was composed of two constituents; one was slightly active by itself, the other was inactive by itself, but greatly augmented the action of the first. For example, hydrochloric acid can hydrolyse protein to some extent, but pepsin greatly increases its action, thus the enzyme was a compound of pepsin and hydrochloric acid; Mn can catalyse the oxidation of diphenols, but Mn + a certain protein is infinitely better. It is interesting that coenzyme functions were ascribed to the

1

2

1 A. Payen, 1795–1871
2 J.-F. Persoz, 1805–68
3 E. Duclaux, 1840–1904

3   *facing page 22*

4

5

6

4   T. Schwann, 1810–82
5   J. J. Berzelius, 1779–1848
6   J. Mayow, 1643–79

7

8

7   M. Berthelot, 1827–1907

8   E. Buchner, 1860–1917

9   M. Traube, 1826–94

9

10

11

12

10 R. Willstätter, 1872–1942
11 J. B. Sumner, 1887–1955
12 E. Fischer, 1852–1919

13

14

13 L. Michaelis, 1875–1949
14 A. Harden, 1865–1940
15 A. Bach, 1857–1946

15

16

17

16 O. Warburg, 1883–
17 H. Wieland, 1877–1957
18 D. Keilin, 1887–1963

The author and publisher gratefully
acknowledge permission to reproduce
the following photographs: Berzelius,
Fischer and Traube, from W. M. Bayliss
Principles of general physiology, Longmans,
Green and Co., London (1918);
Buchner, Harden, Wieland and
Willstätter, from MacCallam and Taylor
The Nobel-prize-winners, Central European
Times Publishing Company Zurich
(1938); Sumner, from Boyer, Hardy and
Myrbäck (1959) *The enzymes*,
© Academic Press 1959,

18

enzyme, and what we should now call the inorganic activator or cofactor was regarded as the catalyst.

There were many other confusions before the true mode of action of enzymes was cleared up. For example in 1902 Schardinger found in milk an enzyme which we now know catalyses the oxidation of aldehyde by the oxidizing dye methylene blue; in other words the aldehyde reduces the methylene blue. But he regarded the aldehyde as the catalyst, enabling the *enzyme* to reduce the dye.

The discovery of the true nature of enzymes had to await their preparation in the pure state, and it took a long time to work out suitable methods. The difficulties were very great. Vernon, in his book on intracellular enzymes in 1908, said definitely

The enzymes are such extremely unstable bodies that it is impossible to purify them without destroying a great deal of their activity... The ideal method would be to follow that used by Mme Curie in isolating radium from pitchblende, when the unknown element was traced by its activity, and by various purification processes obtained in greater and greater concentration, till the preparation of maximum radio-active power was found to be the pure radium salt. Probably such a method would be impossible with enzymes, unless a temporary stability could be artificially induced.

Nevertheless, that is just what we now do.

The serious purification of enzymes did not begin until about 1922–8, when Willstätter undertook the purification of saccharase and one or two other enzymes. Although he did not succeed in obtaining the pure enzymes, he made some progress and developed adsorption methods which were of much value later. He believed that the enzymes were polysaccharides.

I think the first serious attempt with an animal enzyme was probably our own partial purification of xanthine oxidase (later identified with the Schardinger enzyme mentioned above) in 1924–8. It now appears that our best preparations were about one-third pure, a fairly high degree of purification for that time.

This kind of work met with a great deal of opposition and criticism at the time. It was said that the work was 'unphysiological', that the preparations were the products of damaged cells, and that only studies on undamaged cells were significant. The purified preparations were referred to as 'precipitates which behave quite differently from the living cell'. Nowadays the emphasis is all the other way, and to work with pure crystalline enzymes has become something of a status symbol.

The first crystallization of an enzyme was that of jack bean urease by Sumner in 1926, and at first it aroused a good deal of scepticism. It was thought that the enzyme was a mere impurity, adsorbed on crystals of some inert protein. Actually this might well have been the case, as it is now known that the first crystals contained other proteins as well, and it was fortunate that the major component was in fact the enzyme. This achievement was soon followed by the classical isolation of several crystalline proteolytic enzymes by Northrop's group (Northrop *et al.*, 1948), but for many years thereafter the number of purified enzymes was very small, and the great majority of pure crystalline enzymes have been obtained only in the last two decades.

Northrop produced conclusive evidence that his crystals were in fact crystals of the enzymes themselves, and all the enzymes which have been purified since, without exception, have turned out to be proteins. A small number of claims to have found non-protein enzymes have not survived critical examination. So the conception of an enzyme has undergone a development, first from a vague influence or property in certain preparations to a definite chemical substance, and finally to a specific protein.

There was also a change in ideas about the general nature of catalysis by enzymes. The idea that enzymes were surface catalysts was due to the misleading analogy with the inorganic catalysts used by the chemists, such as platinum black or finely divided nickel, the surfaces of which were dotted all over with active centres or irregularities in the structure at which reaction could occur. It was only much later, when methods had been developed for estimating the molecular weights and number of active centres of enzymes that it became clear that in a great number of cases there is only one active centre in each enzyme protein molecule, or sometimes two. Occasionally four subunits, each with one active centre, may combine to form a larger molecule. In short, enzymes are not surface catalysts so much as definite chemical substances which themselves enter into the reaction. The reaction takes place at a point in the enzyme molecule where there is a structure adapted to combine specifically with the substance acted upon (the substrate).

Proteins are extremely complex structures; an average enzyme may have a molecular weight of the order of 100,000, and little could be done to determine the structure of the main bulk of the molecule until special methods had been developed for the purpose, which occurred only recently. Meanwhile, however, it was possible to make a few deductions about the structure of the active centres from two kinds of observations on factors affecting the catalytic activity, namely specificity of action and direct attack by chemical reagents.

Towards the end of the nineteenth century Emil Fischer had greatly added to knowledge of the structure of carbohydrates and proteins, many of which acted as substrates of enzymes. He was thus able to study the range of action or 'specificity' of some enzymes, especially of the saccharase type; and although he had no pure enzymes to work with, he showed that enzymes have a very high degree of specificity for their own particular substrates. This led him to suggest in 1894 that the enzyme and the substrate on which it acts are so constructed as to fit together like a lock and key. This has since been shown to be true to a degree which he could hardly have suspected. Not only the chemical structure but also the shape of the substrate molecule is important.

These observations have been greatly extended by the study of substances closely related in structure to the substrate. Many of these are so similar to the substrate that they may combine with the active site of the enzyme, even though no reaction may follow. These are referred to as 'competitive inhibitors' because they inhibit the action of the enzyme on the substrate by impeding the access of the latter to the active site, and because they therefore compete with the substrate for the site. By determining the effect of making chemical modifications in their structure on their affinity for the active site it is often possible to deduce what chemical groups are involved in the binding and to make inferences about the corresponding structure of the binding-site itself.

Several chemical reagents are available which specifically attack particular kinds of groups in proteins, and if the enzyme in question has such a group in its active centre, the reagent will inactivate the enzyme. The best known of these reagents are those which attack thiol (-SH) groups, and by their use many (but by no means all) enzymes have been shown to depend on such groups for their activity.

In the majority of cases the active centres of enzymes do not contain any special constituent, and they consist simply of patterns of amino-acid residues arranged in such a configuration as to fit the substrate molecule. In many cases, however, the centres may contain other substances, not found in ordinary proteins, such as flavin or pyridoxal or haem groups. Also it is not uncommon for them to contain metal atoms. Thus even before it became possible to determine the chemical structures of proteins, a certain amount had been found out about the structures of the active centres of enzymes.

The complete determination of the chemical structure of the whole molecule of an enzyme is a very difficult matter, and has only very recently been achieved, and only for a very small number of enzymes, notably ribonuclease, chymotrypsin and lysozyme. It requires the patient application of the methods developed by Sanger (1952) at Cambridge for the determination of the order of arrangement of the different amino-acids along the polypeptide chains. But even this is not enough; the chemical structure by itself does not reveal the reason for the catalytic properties of the enzyme. It is necessary to determine how the polypeptide chains are folded up in the molecule so as to produce the active centre with the correct arrangement of groups to fit the substrate and activate it. This can only be achieved by X-ray methods, and it has been done for only three or four enzymes. In 1966 Phillips et al. constructed a complete three-dimensional model of the lysozyme molecule, showing the structure of the active centre and the exact way in which the substrate molecule combines with it, and giving already an indication of the cause of the catalysis. This is truly a major landmark in the development of enzymology.

Kinetics and mode of action

The mode of action of enzymes has largely been studied in conjunction with their kinetics. The subject of enzyme kinetics may be said to begin with the observation of A. J. Brown of the British School of Malting and Brewing, Birmingham, in 1902, that the rate of hydrolysis of cane sugar by saccharase was independent of the sugar concentration over a wide range, although it depended on the enzyme concentration. Brown correctly concluded that the enzyme and the sugar first form a compound, which then breaks down at a certain rate into enzyme and products. If the concentration of the substrate (sugar) is high enough to saturate the enzyme, i.e. to convert all the enzyme into enzyme-substrate compound, addition of more sugar cannot produce any increase in rate. This brings out the essential difference between enzyme reactions and ordinary chemical reactions; namely that in ordinary reactions the rate is proportional to the concentration of the reactant, in accordance with the law of mass action, but in enzyme reactions the rate increases to a limiting value as the reactant concentration is increased to a point at which the enzyme becomes saturated, after which it does not increase further. In the same year V. Henri reached similar conclusions, working in Paris on the same enzyme independently, and he worked out a mathematical theory.

More than ten years later Michaelis (1913) revived the theory of Brown and Henri in rather more detail. Thereafter it became known as the Michaelis theory, and the parts played by Brown and Henri were largely forgotten. The importance of the theory was twofold: it showed the essential difference between enzyme kinetics and those of homogeneous chemical reactions, and it suggested means of determining the affinity of enzymes for their substrates. If the initial velocity of an enzyme reaction is plotted against the substrate concentration, the curve obtained is part of a rectangular hyperbola, in agreement with the theory.

This curve is determined by two parameters, namely the maximum or saturation velocity (V) and a constant known as the Michaelis constant (K_m), which is the substrate concentration at which the velocity is half V. These constants may be evaluated from the experimental curve, but a more satisfactory method was developed by Woolf at Cambridge in 1932. He pointed out that if the reciprocal of the velocity is plotted against the reciprocal of the substrate concentration a straight line is obtained, and that this cuts the two axes at the reciprocal of V and the reciprocal of K_m respectively. His contribution was overlooked, however, and two years later the method was again published by Lineweaver and Burk and has since been known as the Lineweaver–Burk plot. It is extensively used.

An enzyme reaction involving one substrate may be written in accordance with the Michaelis theory as follows:

$$\text{E} + \text{S} \underset{k_{-1}}{\overset{k_{+1}}{\rightleftharpoons}} \text{ES} \overset{k_{+2}}{\rightarrow} \text{E} + \text{P},$$

where the rate constants for the separate steps are as shown. From the theory it may be shown that, provided k_{+2} is small enough for equilibrium to be maintained in the first stage, $K_m = k_{-1}/k_{+1}$, in other words K_m is the reciprocal of the affinity of the enzyme for the substrate. For many years it was generally assumed that this condition was fulfilled. In 1925, however, Michaelis kinetics were superseded by Briggs–Haldane kinetics. Briggs and Haldane showed that if k_{+2} is not small, so that steady-state rather than equilibrium conditions apply to the first stage, K_m is given by $(k_{-1} + k_{+2})/k_{+1}$. K_m is then partly kinetic in nature and is no longer determined solely by affinity. Michaelis kinetics are therefore a limiting case of Briggs–Haldane kinetics.

Briggs–Haldane kinetics held the field until quite recently, but since they only applied to enzyme reactions involving one substrate, whereas the majority of enzymes catalyse reactions between two substrates, they have now had to give way to a variety of more complicated expressions (Dixon & Webb, 1964).

The history of enzymes

Systems

The discovery and separation of enzymes is still proceeding very actively, but in recent years the reverse process has become increasingly important, namely the reconstruction of biological systems by putting together purified enzymes. This is the natural way of studying any complex mechanism, by taking it to pieces, examining the separate parts, then putting them together to see how they co-operate, and finally reconstructing the whole system. Biochemical studies have shown that metabolism consists of a number of chains of successive enzyme reactions, and it is possible to reconstruct several of these multi-enzyme systems from purified enzymes and coenzymes. This has given a great deal of insight into their working.

To reconstruct such systems enzymes alone are usually not enough; coenzymes are also required, in order to act as functional links between one enzyme and another. The majority of coenzymes are specific substances of molecular weight less than 1,000, but of fairly complicated chemical structure, which have the power of acting as carriers of particular chemical groups, taking up the group in a reaction catalysed by one enzyme and giving it up to another molecule in a reaction catalysed by a different enzyme. The majority of enzyme reactions are transfer reactions, in which some group is transferred by the enzyme from combination with one molecule to combination with another. The action of a coenzyme functioning as a carrier of a group X may be written as follows:

$$AX + C \overset{E_1}{=} A + CX, \qquad CX + B \overset{E_2}{=} C + BX$$

where the coenzyme C acts as a link between one enzyme E_1 and another E_2 by receiving the group X from the substrate molecule AX in the first reaction and giving it up again to substrate B in the second, the net result being a transfer of X from A to B.

The first coenzyme carriers found were involved in biological oxidations as hydrogen-carriers. Here X would represent two hydrogen atoms, so that the substrate AX becomes oxidized by the acceptor B, which becomes reduced. The name 'coferment' or 'coenzyme' was introduced by Bertrand in 1897, but the first clear case was the classical observation of Harden and Young in 1906 that the enzymes of yeast juice could be freed by ultrafiltration from smaller molecules, and could then no longer ferment; but on adding the filtrate the power of catalysing fermentation was restored. Actually, however, their 'cozymase' seems to have been a mixture of several coenzymes, and these were not separated or their mechanism understood. About 1915 Batelli and Stern extracted a 'coenzyme of respiration' of animal tissues, which they called 'pnein'; this attracted attention for some years, but was probably much the same mixture of cofactors and substrates as Harden and Young's.

The discovery of the main hydrogen carriers involved in respiration took place after 1921, and will be dealt with in the next section, but there are specific coenzymes carrying many other kinds of groups. For example, the carrier of phosphate groups, adenosine triphosphate (discovered in 1929); the carrier of acyl groups, coenzyme A (1947); that of formaldehyde or formyl groups, tetrahydrofolate (1947); that of methyl groups, vitamin B_{12}; that of carboxyl groups, biotin (1959); that of acetaldehyde groups, thiamine pyrophosphate (1937); that of amine groups, pyridoxal phosphate (1944); and so on. All these coenzymes act as links between enzymes, so enabling systems to be built up catalysing many of the lines of metabolism now known to us.

HISTORY OF BIOLOGICAL OXIDATION*

Having just mentioned hydrogen carriers, which play an important part in biological oxidation and respiration, let us go back in time to the early history of that part of the subject.

I have already mentioned Mayow (1674), and I need only point out that what he referred to as nitro-aerial particles of the atmosphere were really oxygen molecules: his words were written exactly a century before the discovery of oxygen by Priestley. It is hardly necessary to go into the work of Priestley

---

* This lecture was delivered in 1960. A rather fuller account covering much the same field has recently been written by Keilin, and appeared in his book *The history of cell-respiration and cytochrome*, published posthumously in 1966.

in demonstrating that oxygen is used up in respiration, or the work of Lavoisier at about the same time, showing that the phlogiston theory was incorrect, and also that respiration produced carbon dioxide.

The next landmark was the discovery of ozone by Schönbein in 1840. Schönbein was for a time a schoolmaster in England, but did the most important part of his work when he was living in Basel. He was a great friend of Liebig, and was an extremely prolific worker, publishing altogether nearly 400 papers. He was particularly interested in active forms of oxygen in connexion with biological oxidation processes, and at one time did a good deal of work on the two enzymes acting on hydrogen peroxide, which we now call catalase and peroxidase (the names are much later). He pointed out that their action was very similar to that of iron salts, though it was not until long afterwards that it was shown that they do in fact contain iron, being haemoproteins.

The conversion of molecular oxygen, which is rather inert as an oxidising agent, into the strong oxidizer ozone was regarded by Schönbein as the key to biological oxidations. He said in 1857 'As a matter of fact, without the presence of such substances as convert ordinary oxygen into ozone, animals would be suffocated in the midst of an ocean of the purest but inactive oxygen as quickly as in a vacuous space'.

This view led to a general 'ozone craze' for many years. To quote Kastle,

During the years 1866–7 daily observations were made on the amount of ozone in the atmosphere of Paris and other localities, and everywhere men were busily engaged in studying its relation to health and disease...According to Dr Moffat, an English physician, the approach of an ozone period, during which the quantity of ozone in the atmosphere suffers a considerable increase, is followed by a corresponding increase in the luminosity of the glowworm and in man by an increased phosphate excretion, as well as by the approach of thunderstorms, and a marked increase in the number of cases of toothache, neuralgia, apoplexy etc...The most exact agreement was established between the warnings of the 'Admiralty cautionary telegrams', as the British weather forecasts were then called, and the readings of the Doctor's ozonometer.

To this day, holiday resorts are sometimes praised for their ozone, as a direct consequence of this theory of biological oxidations.

Ozone, however, will not in fact carry out biological oxidations, and is actually a toxic substance.

The next stage was the antozone theory. According to Schönbein, oxygen underwent the change

$$O \rightarrow \underset{\text{ozone}}{\ominus} + \underset{\text{antozone}}{\oplus}$$

Antozone combined with water to form hydrogen peroxide, another form of active oxygen. This gave rise to a 'peroxide era', which dominated the subject until about 1920, with a whole series of theories involving various kinds of peroxides, organic peroxides, holoxides, moloxides, etc., formed by direct combination of the oxygen molecule with some organic substance:

$$
A + \quad \overset{\text{O}}{\underset{\text{O}}{\|}} \quad \longrightarrow \quad A \overset{\text{O}}{\underset{\text{O}}{<|}}
$$

The peroxide was then supposed to bring about the oxidation of organic substances. This idea was supported by the discovery of the enzyme peroxidase, which was shown to catalyse the oxidation of certain organic substances by hydrogen peroxide.

Two other theories of biological oxidation which did not involve peroxides were put forward a little later. The theory of Clausius (about 1860) was that the oxidations were done by atomic oxygen, formed by dissociation of the oxygen molecule. The theory of Hoppe-Seyler (1878) agreed that the oxidations were due to atomic oxygen, but assumed that this was formed from the oxygen molecule not by dissociation but by reduction:

$$2H + O_2 = H_2O + O$$

These theories, however, did not seriously interfere with the general acceptance of the peroxide idea.

This was given fresh impetus by the work of A. Bach, who was prominent in this field for many years, and to whom we owe the name peroxidase. In 1903 Bach and Chodat published a classical paper which largely dominated the subject up to about 1920, though in fact it proved to be misleading in some respects. Mainly on the basis of experiments on preparations from plant tissues, they maintained that the oxidation system could be separated into two parts; one they named oxygenase and the other peroxidase. Oxygenase converted oxygen into an organic peroxide, and peroxidase used this for the oxidation of other substances. It was not until more than twenty years later that the nature of their system became clear and it was seen that in two essential points it did not behave as they had supposed. What had been called oxygenase (the term is now used with a different meaning) was in fact the enzyme now known as catechol oxidase, which oxidizes catechol and similar substances by means of molecular oxygen. The oxidizing substance formed by the oxygenase preparation was not a peroxide but a quinone, the product of the oxidation of a catechol substance contained in the preparation. Moreover the peroxidase did not play the part assigned to it in the system, for peroxidase does not make use of quinone as an oxidizing agent. It had been commonly supposed that the oxygenase–peroxidase mechanism would account for the respiration of animal tissues as well as plant tissues, but catechol oxidase is characteristically an enzyme of plants.

Owing to the concentration of attention on oxygen at that time, it does not seem to have been clearly realized that the oxidation of one substance automatically involved the reduction of another, so that an oxidizing enzyme must also be a reducing enzyme. It was known, however, that reducing enzymes did exist (for example the Schardinger enzyme already mentioned), and Bach put forward an analogous theory to account for their action. This was that, just as the oxidizing enzyme consisted of peroxide and peroxidase, the reducing enzyme consisted of a perhydride and a perhydridase. No evidence has been obtained of the presence of either of these hypothetical components.

Although none of the above theories is now held, it may be mentioned that hydrogen peroxide can be formed from oxygen by some enzymes, but by the direct reduction of the oxygen molecule

$$A\!\!\!\begin{array}{c} H \\ \\ H \end{array} + \begin{array}{c} O \\ \| \\ O \end{array} \longrightarrow A + \begin{array}{c} HO \\ | \\ HO \end{array}$$

as suggested by Wieland and first shown in our laboratory in 1925 for the case of the Schardinger enzyme. It is now known to be a general property of flavoprotein oxidases.

The last of the oxygen-activation theories was that developed by O. Warburg in Berlin from about 1920 to 1930, which was that the oxygen molecule was first activated by combination with an iron atom in a special haemoprotein enzyme, 'the respiratory ferment'. Unlike the earlier theories, this was based on much solid experimental evidence obtained with the actual respiration systems of yeast, bacteria and animal tissues. Some elegant physical methods were devised for the purpose, and the properties of the enzyme were established by actual measurements on the respiration itself. The respiratory ferment was later identified by Keilin with one of the cytochromes (now known as cytochrome a_3 or cytochrome oxidase).

All the theories mentioned up to now have been essentially oxygen-activation theories, that is to say they have been possible mechanisms for the conversion of the somewhat inert oxygen molecule into some actively oxidizing form. In about 1920 there came a complete revolution of thought, with the so-called hydrogen-activation theory of Wieland. This was briefly that it was not the oxygen molecule that needed activation so much as the organic molecules which were to be oxidized. They were inert towards molecular oxygen and other mild oxidizing agents, but much evidence was produced that they could be activated by certain chemical catalysts in such a way that two hydrogen atoms within the molecule became more loosely attached and could readily be transferred to oxidizing agents. It was pointed out that most biological oxidations are in fact dehydrogenations rather than additions of oxygen atoms. The biological importance of this was seen when in 1920 Thunberg discovered in animal and plant tissues a whole group of enzymes with exactly this function (the dehydrogenases). These enzymes are now known to be a most essential part of the biological oxidation system, and their importance can be gauged from the fact that at the moment well over 150 different dehydrogenases are known, activating different substances.

The position was therefore that at one extreme we had oxygen activation and at the other hydrogen activation (or more truly substrate activation). The next phase was the discovery of a number of intermediate carriers which linked the two by becoming alternately reduced by one system and oxidized by the other. The first suggestion of such a hydrogen-carrier action was probably made by Hopkins in 1921 in connexion with the sulphur-containing tripeptide glutathione which he had just discovered. It is true that glutathione does undergo oxidation and reduction in tissues, and enzymes involved in the process are known to exist; it must therefore act as a hydrogen-carrier to some extent, but it is still not clear whether this makes any significant contribution to the total respiration.

An important landmark was the discovery of the cytochromes by Keilin in 1925 by spectroscopic observations. The oxidation and reduction of these haemoproteins can be followed by the spectroscope even while they are in the respiring tissue, and by rate measurements Keilin was able to show that in the tissues tested a large part of the respiration was due to this process. In many tissues five different cytochromes (known as b, c_1, c, a and a_3) act together in respiration, but in series, not in parallel. With the exception of a_3, none of them reacts with oxygen; they therefore form a chain of carriers in the order shown, each one reducing the one on the right of it until the last one reduces oxygen. By the use of specific inhibitors, Keilin was able to identify this last cytochrome (a_3) with Warburg's respiratory ferment. These cytochromes are not capable of reacting directly with the dehydrogenases, and still other hydrogen-carriers come between.

The next landmark was the discovery and elucidation of the mode of action of the first of these other carriers, reacting directly with the dehydrogenases, by Warburg in 1931–6. This is the pyridine-nucleotide coenzyme, nicotinamide-adenine dinucleotide (NAD), or for some dehydrogenases its phosphate (NADP). Each reacts with a large number of different dehydrogenases, becoming reduced to its dihydro-form. The reduced form is reoxidized by the next carrier in the chain, a flavoprotein enzyme, composed of a protein part combined with a reducible flavin group. The first flavoprotein was discovered and studied by Warburg in 1932. The flavoprotein in turn is reoxidized by the remaining carrier, ubiquinone, a quinone discovered in 1953, which links up with the first of the cytochromes, so completing the chain.

Thus, far from having a single hydrogen-carrier to link the activated substrates with the activated oxygen, the respiratory system has a chain of at least eight successive carriers to perform this function. The discovery and study of these intermediate carriers has constituted an important phase of the development of the subject, extending over about forty years almost to the present day.

In recent years the purpose of this complicated arrangement has been traced to the necessity of the efficient utilization of the energy available from the oxidation for the biological processes requiring energy, which is the main function of the whole system. The energy available from the reaction between an average substrate and oxygen is enough for the formation of several 'energy-rich bonds' (the form in which energy is stored for biological purposes). If the reaction occurred in a single step, only one such bond could be generated, but by splitting it up into a series of stages, several of the steps would have enough energy to form one bond each, so that several such bonds could be formed by the overall reaction. The present phase of development of the subject is largely concerned with the investigation of the mechanisms by which this is brought about. But at present this can scarcely be called history.

One further important point must be mentioned, namely that this oxidation system is located in small intracellular particles, the mitochondria, and is not in free solution within the living cell. Mitochondria have long been known to the histologists, but

The history of enzymes

there was no indication of their function until in 1912 Kingsbury suggested that they might be the sites of cellular oxidations, and in the following year Warburg showed that respiration was associated with the granular part of the cell structure. It was not possible, however, to prepare these particles until about 1948, when the advent of differential high-speed centrifugation gave a means of isolating mitochondria in an apparently undamaged condition. It was then found that the whole of the cytochrome system described above is located within these particles, and that they respire in much the same way as the complete cells. Mitochondria and their enzyme systems are being intensively studied at the present time. They have a complex structure, and the relation between their structure and function is proving to be of great interest.

REFERENCES

Berzelius, J. J. (1837). *Lehrbuch d. Chem.* Dresden and Leipzig: Arnoldischen Buchhandlung. 3rd ed. **6,** 19–25.

Dixon, M. & Webb, E. C. (1964). *Enzymes.* London: Longmans. 2nd ed. pp. 92–116.

Keilin, D. (1966). *The history of cell-respiration and cytochrome.* Cambridge University Press.

Liebig, J. (1839). *Liebig's Ann.* **30,** 250.

Northrop, J. H., Kunitz, M. & Herriott, R. M. (1948). *Crystalline enzymes.* Columbia University Press.

Payen, A. & Persoz, J. F. (1833). *Ann. Chim. (Phys.)* **53,** 73.

Phillips, D. C. *et al.* (1967). *Proc. Roy. Soc. Lond.* B **167,** 365–388.

Sanger, F. (1952). *Adv. Protein Chem.* **7,** 1.

3

THE DEVELOPMENT OF
MICROBIOLOGY

by E. F. Gale

Micro-organisms make their major impact on man because they can cause disease. The first mention of the possibility that diseases could be due to sub-visible creatures was made about 1546 by Fracastorius when teaching in Padua University. He concerned himself with the nature of contagion, and similarities in the spread of diseases in animals and plants led him to suggest that the contagious nature of such diseases could be due to the presence in the air of 'seminaria', germs or seeds. This was, of course, hypothesis and no serious study of micro-organisms was possible until their existence could be demonstrated. The first observations of micro-organisms were made by the Dutchman, Leeuwenhoek (1632–1723), a linen draper and book-keeper of Delft who held various public offices such as Chamberlain of the Council Chamber of the Sheriffs of Delft. In his spare time he took up the grinding of lenses as a hobby. Until the time of Leeuwenhoek crude lenses had been made by such devices as allowing drops of molten glass to cool in mercury and using suitable drops as magnifying glasses. The distortion and poor resolution obtained made such lenses suitable only for the study of larger creatures. Leeuwenhoek discovered how to grind lenses to produce optical surfaces with accurate curvature and the products enabled him to observe material and organisms invisible to the naked eye. It is possible to calculate from such of Leeuwenhoek's lenses that still exist that he obtained magnifications of 175–250 diameters with some of his better preparations. He used these lenses to study the structure of such things as the mouth and eye of the bee and louse, and to observe the detail of various moulds and fungi. His observations were written up in a series of letters submitted to the Royal Society in London beginning in 1673. These letters caused great interest amongst the Fellows of the Royal Society, who elected Leeuwenhoek a Fellow in 1680, although he never came to London to be received into the Fellowship. The letters were written in a somewhat difficult Dutch and the extent of our knowledge of Leeuwenhoek's observations is due to the excellent translations by Clifford Dobell in his book *Leeuwenhoek and his little animals.*

The first descriptions of true bacteria are set out in a letter written by Leeuwenhoek in 1676. At this time he was examining such things as infusions of pepper, water from ponds and streams, saliva and scrapings of teeth. From his descriptions and drawings it is possible to recognise algae, rotifers, ciliates, protozoa and bacteria. According to Dobell's translation, bacteria were first described in the following way: 'the fourth sort of little animals were incredibly small, nay, so small in my sight that I judged that even if 100 of these very small animals lay stretched out one against another, they would not reach to the length of a grain of coarse sand.' Leeuwenhoek was intrigued not only by the minuteness of these little animals but also by the very large numbers that he found in drops of pond water or pepper infusion; he remarked that the number of micro-organisms that he could count in a drop of water exceeded the human population of the world. Once he had discovered micro-organisms he began to search for different sorts of organisms in all sorts of places and sources. The story is told that he was awakened one night by the noise of a burglar entering his house. He got out of bed, went quietly downstairs and found the burglar in his living room. He seized the burglar by the hair, pulled back his head and then, to the intense surprise of the very frightened burglar, pulled out a knife and took scrapings of the man's green and decayed teeth. Here, thought Leeuwenhoek, was an excellent source of new organisms.

Leeuwenhoek's 'microscopes' were not microscopes as we know them today, but were single lenses held against the eye and used to observe objects fixed on a needle close to the front of the lens (see plate 1). It is believed that he made over 400

39

lenses, some of which were deposited with the Royal Society and some in the museums in Holland. Unfortunately the lenses which were deposited at the Royal Society in London have been lost. The precise manner in which Leeuwenhoek used his lenses is not understood today. There must have been some

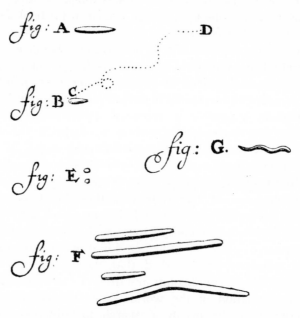

Fig. 1. Leeuwenhoek's figures of bacteria from the human mouth
(from *Antony van Leeuwenhoek and his 'little animals'*, by Antony van Leeuwenhoek, trans. and ed. by C. Dobell, Dover Publications Inc., New York)

trick of illumination which made possible the accurate observation of micro-organisms. Although Leeuwenhoek allowed visitors to look through the lenses to see the marvels that he described, he was never willing to allow his visitors to see him set up the microscopes before use. In fact he says in one of his letters, 'my method for seeing the very smallest animalcules and eels I do not impart to others . . . that I keep for myself alone'. Although kings, princes, clerics and professors flocked to Delft to see the little animals, these persons were only allowed to look through his microscopes by appointment and after the apparatus had been set up. Nevertheless it is clear from the detailed descriptions and drawings in the letters that Leeuwenhoek was able to use his lenses to observe bacteria. Further, he described the main morphological types that we recognize today, observed motility in certain organisms and noted the effect of heat in stopping motility and decreasing the numbers of micro-organisms. For example, he observed that the number of organisms in the saliva was decreased after drinking a cup of hot tea.

Leeuwenhoek described bacteria in 1676; it was nearly 100 years before bacteria were again reliably recorded and they were not studied in any detail until 200 years had passed. The development of microbiology from that time centred on three controversies; spontaneous generation, the nature of fermentation, and the germ theory of disease. The word 'microbe' itself was first introduced by Sédillot in 1878 and used to designate any organism so small as to be invisible to the naked eye but visible under the microscope, and in 1882 Pasteur suggested that the science of microbial life should be called 'microbie' or 'microbiologie'.

SPONTANEOUS GENERATION

By the time of Leeuwenhoek, the earlier belief in spontaneous generation of the larger forms of life such as bees from lions, mice from dirt or eels from mud, had been largely discarded. However, Leeuwenhoek's observation of micro-organisms and their wide distribution reopened the discussion on the possibility that living creatures could arise spontaneously. Modern sterilisation techniques have largely developed from experiments designed to prove or disprove the spontaneous generation of micro-organisms. The difficulties in disproving spontaneous generation included such factors as the universal distribution of micro-organisms—with the consequent ease of contamination of any apparatus used for experiments—the presence of heat-resistant forms of micro-organisms, especially spore-bearing organisms, and the ability of some species of micro-organism to grow in the absence of air. Typical of the early experiments were those carried out in 1711 by Joblot who boiled infusions of hay in tubes and then covered the tubes with parchment. The infusions remained clear and sterile until the tubes were opened to the air when 'animals developed from eggs dispersed in the air'. The experiments occasionally failed, and later attempts involved essentially the same principles with greater care and safeguards taken against accidental contamination. Some of the most successful experiments along these lines were carried out by Spallanzani (1729–99) who boiled infusions of hay in flasks and then sealed the flasks. In the vast majority of cases the contents of the flasks remained free from micro-organisms until the necks of the flasks were broken and their contents exposed to the air, when micro-organisms soon appeared and grew in the infusions. Spallanzani studied these organisms under the microscope and was the first to show that bacteria grow and multiply by fission. This involved the use of an ingenious trick: he took a drop of the infected infusion on a slide and observed it under the microscope; alongside the drop of infected infusion he placed a drop of sterile infusion and joined the two drops by a narrow trace of liquid. He then watched the bacteria in the infected drop and waited until a single cell moved down the trace and into the sterile drop. At this point he broke the joining trace so that he had one organism in the drop of sterile medium. He was then able to observe the growth and eventual division and multiplication of the single cell in the sterile drop. These experiments showed that microbial growth could be prevented by heating the liquid infusions, that no growth occurred unless organisms from the air contaminated the infusion, and that micro-organisms grew from micro-organisms. Observations of this nature led directly to the development in 1810 by Appert of a commercial process for the preservation of fruits, juices and meats by placing these materials in clean bottles, corking them thoroughly and raising their temperature to 100 °C.

Spallanzani and his colleagues claimed that their experiments demonstrated that micro-organisms did not and could not arise spontaneously. Their opponents claimed that the experiments succeeded either because air was excluded from the

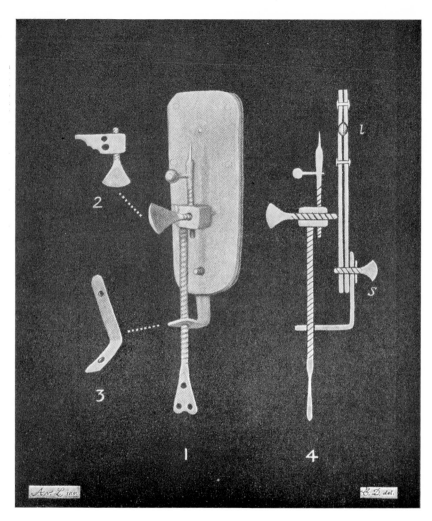

Plate 1. Leeuwenhoek's microscope
(From *Antony van Leeuwenhoek and his 'little animals'* by Antony van Leeuwenhoek,
trans. and ed. by C. Dobell, Dover Publications Inc., New York)

facing page 42

vessels and was necessary for the growth of the organisms or, alternatively, that heating of the infusions vitiated the media so that nothing could grow in them anyway. Gay-Lussac suggested that the success of Appert's commercial preservation process was due entirely to the exclusion of air from the vessels. The criticism had to be answered and so it was necessary to repeat the experiments in a way that would allow access of air, which required in turn that the air must be scrupulously cleansed from micro-organisms before entering the vessels. Early experiments along these lines involved heating infusions in vessels which were connected to the atmosphere through wash bottles containing sterilizing liquids such as concentrated sulphuric acid or caustic potash. Schulze was successful in demonstrating that under these conditions it was possible to sterilize infusions and maintain them sterile although in contact with air. Experiments were made much less difficult after Schroder and Von Dusch showed in 1854 that cotton-wool could be used as a filter for micro-organisms. There were, nevertheless, occasional failures in all these experiments and organisms grew up in the media despite all precautions. These failures were particularly frequent when the liquid medium contained either milk or meat extract.

The controversy was finally settled by Pasteur (1822–95) who succeeded by closing the loop holes in the argument by better technique and greater attention to experimental detail. He said 'Every source of error plays into the hands of my opponents. For me, affirming as I do that there are no spontaneous fermentations, I am bound to eliminate every cause of error, every perturbing influence. Whereas I can maintain my results only by means of the most irreproachable technique, their claims profit by every inadequate experiment.' The Franklands, contemporaries of Pasteur, writing in 1898 said 'Pasteur was not only a savant content to seek the truth, but when he had in any matter succeeded in the difficult task of convincing himself, he was impelled with almost a fanatic's zeal to force his conviction on the world, nor did he put up his sword until every redoubt of unbelief had been taken, every opponent converted or slain.' The classical experiments carried out by Pasteur

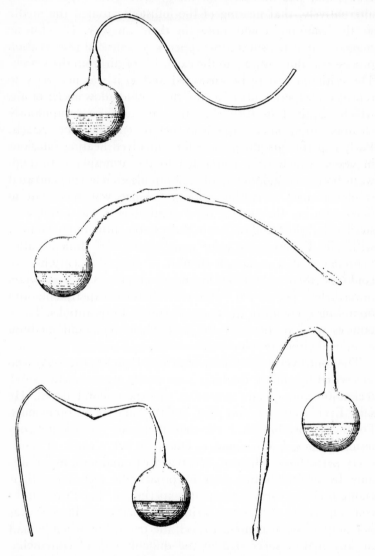

Fig. 2. Swan-neck and other types of flasks used by Pasteur in experiments to disprove spontaneous generation of microbes (from *Oeuvres de Pasteur*, Vol. II, *Fermentations et generations dites spontanées*, Masson, Paris, 1922)

involved the use of swan-neck flasks (see fig. 2). He placed extracts of yeast in these flasks and sterilized them in the usual way by heat; the necks of the flasks were then drawn out into a long S-shaped neck, but the neck was not sealed. As the contents of the flasks cooled, air was drawn into the flask down the swan neck and microbes did not enter the flask. The explanation seems to be that during the cooling of the flasks a certain amount of condensation occurred in the swan neck, the condensation water washed the air as it entered the flask and effectively prevented the ingress of micro-organisms during the cooling process. Once the flask was cooled, the shape of the neck prevented any micro-organisms dropping into the flask but allowed free access of air to the contents. Media sterilized in such flasks remained sterile until the neck of the flask was broken off. When this happened rapid contamination occurred and micro-organisms grew in the medium. It is still possible to see, in the museum of the Pasteur Institute in Paris, flasks sterilised by Pasteur and containing crystal clear sterile liquid. Pasteur demonstrated the existence of micro-organisms in the air by sucking air through a plug of gun cotton, dissolving the gun cotton and showing with microscope or by culture that micro-organisms had been abstracted by the cotton plug from the air. The method led to the finding that samples of air from different localitities contained different sorts of organism; this implied in turn that micro-organisms are contaminants, and not a part of air. Pasteur followed up this idea by sterilising extracts of meat or yeast in flasks and sealing the flasks; under these conditions the contents of the flasks remained sterile indefinitely. Flasks were then taken to different localities and the necks of the flasks broken; again different mixtures of organisms were cultured from different places. Most important; when flasks were opened in the clean, pure air at the top of a mountain some of the flasks always remained sterile. This confirmed Pasteur's conclusions that micro-organisms are simply carried around in the air and that, if it were possible to obtain thoroughly clean air, then no micro-organisms would be found therein. Pasteur's particular opponent in this matter was Pouchet who, although he repeated Pasteur's experiments with sealed flasks, was unable to confirm the results. In fact, Pouchet's flasks always became contaminated and were always unsterile. The explanation lies partly in that Pouchet's technique of sterilization was inferior to that of Pasteur's, and partly in the fact that the two scientists used different types of media. Pouchet used hay infusion, which we now know is likely to be heavily contaminated with heat-resistant spores, whereas Pasteur used extract of yeast as his medium and this is more readily and completely sterilized by the heating procedures then adopted.

Confirmation of the fact that clean air is devoid of microbes was provided by Tyndall by application of the 'Tyndall beam' method of showing the presence of dust in air. In these experiments tubes containing urine, sterilised by heat, were placed in the bottom of a cabinet fitted so that a beam of light passed through the air above the tubes. Under normal conditions the urine was quickly contaminated and grew thick cultures of organisms. If, however, the box was left undisturbed until the beam of light showed that the air was 'optically empty' before the sterile urine was introduced, then the urine remained clear as long as the air above the tubes remained undisturbed. Once the door of the cabinet was opened to admit 'optically dirty' air, then the tubes became contaminated and cultures grew in the urine.

In all these experiments there were the occasional failures, particularly when infusions of hay were used. Pasteur never admitted failures of this nature, but most of the other workers in the field sooner or later obtained cultures of organisms under conditions in which the media had been sterilized by heat and then adequately shielded from dust in the air. Bastian found that urine which had been 'sterilized' by boiling sometimes gave rise to cultures after neutralization. In 1877 Cohn observed that the organisms that were obtained from infusions of hay differed according to whether the material had been heated to 100° C or not, and experiments of this nature led gradually to the appreciation that there are forms of some organisms which are able to survive heating to 100° C. In this manner the existence of heat-resistant spores first became known. Spores can be killed by heating to a temperature of 120° or higher and this is the reason for the modern use of autoclaves for sterilizing by steam under pressure. Tyndall devised another method for sterilizing media even although they contained spores. In 1877 he wrote to Huxley

Before the latent period of any of the germs has been completed, I subject it for a brief interval to a temperature which may be under that of boiling water. Such softened and vivified germs are on the point of passing into active life and are thereby killed, others not yet softened remain intact. I repeat this process well within the interval necessary for the most advanced of those others to finish their period of latency. The number of undestroyed is further diminished by this second heating. After a number of repetitions which varies with the nature of the germs, the infusions, however obstinate, are completely sterilised.

This process of 'Tyndallization' has proved useful for the sterilization of media which contain components which might themselves be damaged or destroyed by heating to high temperatures. For example, media containing milk or blood, which would be denatured by heating to 100° C, can be adequately sterilized by subjecting them to the Tyndallization process on three successive days.

THE NATURE OF FERMENTATION

The biological nature of alcoholic fermentation was established between 1836 and 1837 by the work of three men: Cagniard de Latour, Schwann and Kützing. All three observed the presence of yeast cells in beer vats, established that the cells were living organisms, and that the fermentation was a vital activity dependent upon the presence of the living cells. Cagniard de Latour described yeast cells in detail and also the budding process by which they multiply. He gauged their size as 1/150 mm. Schwann gave the name 'Zuckerpilz' or sugar fungus to the organisms (*Saccharomyces*) and established that (*a*) the fungus was always present during fermentation, (*b*) if the fungus was destroyed by heat or chemicals, then fermentation stopped, and (*c*) that the fermentation 'principle' increased during the process and this increase accompanied the reproduction and multiplication of the yeast. Kutzing observed that the organisms associated with the alcoholic fermentation on the one hand and with the vinegar fermentation on the other seemed to be different.

These ideas were in opposition to chemical thought current at that time, which maintained that both alcoholic and vinegar processes were chemical reactions capable of expression in simple chemical equations, and that the sludge or turbidity that occurred during the processes were either accidents or possibly inanimate catalysts. The detailed description of the organisms found in fermentation vats aroused derision amongst the chemists, a derision that was expressed in an anonymous article appearing in 1839 in Liebig's *Annalen der Chemie*. This described beer yeast as possessing the shape of a tiny distillation flask without the cooling apparatus; having no noticeable teeth and eyes but an easily distinguished stomach, intestinal canal, anus (as a rose coloured point) and urinary secretory organs. The article continued with a description of how the organisms escaped from their eggs, gobbled up the sugar from solution, digested it in the stomach, evacuating alcohol from the intestinal tract, and carbon dioxide from the enormously developed sexual organs.

Pasteur entered the field of fermentation in 1857 with his studies on the nature of the lactic fermentation. He found that the process was dependent on the presence of a grey deposit of an organized nature, and that the cells forming this deposit were considerably smaller than yeast cells. Sterile medium was not fermented but the addition of a small amount of the deposit from a lactic fermentation was followed by lactic fermentation in the medium and an increase in the amount of the deposit. Microscopic observation showed that all the individuals in the deposit were the same.

Pasteur's paper on the lactic fermentation provided the first detailed description of a microbiological fermentation and, at the same time, a firm statement of the fact that there is specificity between the type of fermentation and the organism that causes it. Pasteur described the effect of pH on mixed fermentations showing that, under acid conditions, yeast fermentation predominated whereas under neutral conditions, lactic fermentation took over from yeast fermentation. The paper also included the first description of differential toxicity in that Pasteur showed that onion juice will inhibit the growth of beer yeast while having no effect on the growth of the lactic organisms. The paper shows all the flair for the correct conclusion that permeates Pasteur's investigations, and provides an interesting comment on his philosophy: 'if anyone should say that my conclusions go beyond established fact, I would agree in the sense that I have taken my stand unreservedly in an order of ideas, which, strictly speaking, cannot be irrefutably demonstrated'.

The chemists believed that alcoholic fermentation was a straightforward chemical reaction in which sugar was converted to alcohol and carbon dioxide. If the yeast cells were a part of the process, then they were only necessary in so far as they were responsible for the production of an albuminoid catalyst which itself accelerated the chemical breakdown. Pasteur attacked both of these ideas. In the first place he showed that there were other products of alcoholic fermentation, such as glycerol and succinic acid, and that it was not possible to express the reaction in simple quantitative chemical terms. Secondly he devised a simple synthetic medium consisting of salts (made from incinerated yeast), ammonia and sugar which he inoculated with a minute amount of yeast. The amount of yeast added was far too small to account for any significant amount of albuminoid catalyst but alcoholic fermentation occurred, and the alcoholic fermentation was accompanied by a growth of yeast: the growth of the yeast cells and the alcoholic fermentation were inseparable. The investigation was the first demonstration that a living organism could be cultured on a completely synthetic medium, devoid of organic matter. A parallel demonstration was carried out by Raulin, a pupil of Pasteur's, who produced a completely synthetic mineral medium for the growth of *Aspergillus niger*.

By 1859 Pasteur had devised methods for separating organisms responsible for different types of fermentation and was able to work with 'pure' cultures and selective media. With these cultures and media he was able to produce a given type of fermentation at will. In this way he produced and studied the lactic, tartaric, acetic and butyric fermentations and obtained cultures of organisms specific for each type of fermentation. He found that the organisms associated with the butyric fermentation were motile when examined under the microscope but that the motility of the organisms ceased near the edge of the cover slip. In parallel with this, he observed that the butyric fermentation was arrested by the presence of oxygen, and so made the discovery that life was possible in the absence of oxygen and that some micro-organisms can lead an anaerobic existence. Such organisms were to be found in natural surroundings as a result of anaerobiosis created in the environment by the presence of other, aerobic organisms. He found that alcoholic fermentation was associated with anaerobic life of yeast cells and proposed that fermentation was the means by which life without oxygen was made possible, oxidation under these conditions being accomplished at the expense of fermentable substances.

Pasteur's experiences with the specificity of fermentation processes led him to understand that different organisms carry out different chemical reactions. This experience was later put to good use when he was called upon to investigate the so called 'maladies' of wines, when he found that many of the troubles of the wine industry, such as the occurrence of sourness, acidity, bitterness or ropiness in wines, could be traced to contamination of the wine liquors with foreign organisms. This led him in turn to the development of methods of 'pasteurization' for the elimination of unwanted organisms from natural juices and to the development of methods of control of wine fermentations in general. Pasteur also paid some attention to the brewing process and advised brewers on methods of hygiene in the brewery to prevent their beers from 'going bad' but he never developed interest in the improvement of beer to the extent that he did for wine. It would appear that Pasteur did not appreciate beer to the extent that he did wine. It is, however, recorded that in 1871 Pasteur paid a visit to Whitbread's brewery in London. In an account of the visit:

The development of microbiology

I was allowed to visit one of the large London breweries. As no one there was familiar with the microscopic study of yeast, I asked to perform it in the presence of the managers. My first test dealt with some porter yeast obtained from the outflow of the fermentation vats. One of the disease organisms was found to be very abundant in it. I concluded therefore that the porter was probably unsatisfactory and was told in fact that it had been necessary to obtain that very day a new sample of yeast from another brewery... I then asked to study the yeasts of other beers. I was given two kinds. One was slightly clouded; on examining a drop of it I immediately recognized three or four diseased filaments in the microscope field. The other was almost clear but not brilliant; it contained approximately one filament per field. These findings made me bold enough to state in the presence of the Master brewer, who had been called in, that these beers would rapidly spoil and that they must already be somewhat defective in taste—on which point everyone agreed, although after long hesitation. I attributed this hesitation to the natural reserve of a manufacturer whom one compels to declare that his merchandise is not beyond reproach.

When I returned to the same brewery less than a week later, I learned that the managers had made haste to acquire a microscope and to change all the yeasts that were in operation at the time of my first visit.

The work of the biologists of the mid-nineteenth century thus laid the foundations for our modern understanding of the microbial origin of fermentation and other forms of decomposition of natural materials. It provided the experimental basis which demonstrated specificity between the organism and the nature of the reaction produced and the establishment of this specificity was to prove all-important in the next phase; the establishment of specificity between disease and the organism that causes it. It is interesting to reflect today that Pasteur spent so much time and ingenuity to prove that fermentation was due to the growth of a living cell and not to an albuminoid catalyst, whereas the biochemists of the early twentieth century were concerned with proving that the fermentation process was brought about by a series of protein enzymes that could be extracted from the cells. In the same way the scientists of the twentieth century do not blench at the thought of spontaneous generation, although admittedly on a different time scale and believe that life must have originated by a spontaneous process occurring within the 'primordial soup'.

THE GERM THEORY OF DISEASE

Just as the observation of micro-organisms in fermentation vats led to dispute between those who believed that the micro-organisms caused fermentation and those who believe that they either were there by accident or were products of the fermentation, so the observation of micro-organisms in diseased animals and plants led to controversy between those who maintained that the micro-organisms were the cause of disease and those who maintained that they were either accidents or products of diseased tissue. In 1834 Bassi established that a fungus caused a distinctive disease of silk worms, and in 1846 Berkeley showed that potato blight was constantly associated with the presence of a certain type of fungus. At this time Pasteur was beginning his investigations in the fermentation field, but he was not at first concerned with putrefaction processes although he was clearly interested in the possible causation of disease by micro-organisms as was shown by his use of the word 'maladie' for effects of foreign organisms on wine. His demonstration of specificity in the relationship between fermentation and micro-organism certainly gave rise to the belief that specific diseases could be caused by specific organisms. However, the first practical application of Pasteur's ideas to the field of medicine was made in Glasgow by the surgeon Lister, who realised that putrefaction of wounds could be due to the presence of micro-organisms in those wounds. Lister was deeply interested in Pasteur's findings and decided that, if his ideas were correct, then it should be possible to prevent putrefaction of wounds by preventing the growth of micro-organisms in those wounds. He therefore invented the antiseptic technique of surgery, in which a spray of carbolic acid was maintained over the area of the operation and the wound was covered with carbolic acid dressings in an attempt to kill any organisms that might fall into the wound during surgical procedures. The results on surgical mortality were dramatic, the rate falling from about 80 to less than 10 per cent. and exploratory surgery of the chest and abdomen became feasible. Lister's work established a connexion between wound putrefaction with subsequent 'blood poisoning' and microbial contamination of the wound. The argument that other diseases were likewise due to microbial contamination was an obvious logical step.

To establish specificity between an organism and a disease it was first necessary to devise a method of obtaining bacteria in pure culture. Pasteur had obtained what he thought were pure cultures of fermentation organisms by massive dilution of his cultures to the extent where it seemed probable that his inocula were grown from single cells and were therefore *a priori* pure cultures. It seems doubtful now whether Pasteur obtained really pure cultures of his fermentation organisms by this method; he certainly obtained enriched cultures, cultures enriched to the extent that when he inoculated sugar solutions with them he was able to predict accurately what type of fermentation would result. However, it was the work of Koch (1843–1910) that first provided a means of isolating bacteria in pure culture. Koch invented the method of growing organisms on the surface of media solidified in gelatine. This made possible the modern method of streaking cultures across a solid medium (solidified in agar) until the continuous dilution results in the separation of single cells which then grow as discrete colonies on the surface of the gel. Organisms can then be taken from a single colony on such a 'plate', cultured and the plating procedure repeated until observation indicates that all cells within the culture are of the same type. In this way Koch was able to demonstrate that some of Pasteur's fermentation cultures were, after all, mixtures of different bacteria. The plate method enabled the isolation in truly pure culture of organisms associated with specific diseases. The disease investigated in great detail was anthrax, produced by infection with a spore-bearing aerobic bacillus. Rod-shaped organisms had been observed in anthrax cadavers by Davaine in 1850, but he was unable consistently to reproduce the disease by inoculation or to obtain the organism consistently from anthrax cadavers. In

1876 Koch confirmed the presence of rod-shaped bacteria in all animals dying of anthrax and was able to grow the organisms in blood culture and use the plate method to obtain pure cultures. He published detailed descriptions of the appearance of these organisms under the microscope, described their filamentous form of growth and formation of spores. Cultures were grown in laboratory media, inoculated into experimental animals, and produced anthrax in those animals. Koch observed that the ability to produce anthrax was not a property of all rod-like bacteria; for example, organisms obtained from infusions of hay showed a similar morphology but were unable to produce disease on inoculation into animals. There are numerous rod-shaped bacilli present in the intestinal contents of animals and after death these organisms may spread throughout the tissues of the cadaver and confuse any attempts to isolate the disease-causing agent. In 1840 Henle laid down three conditions which must be satisfied before a given organism can be associated with a given disease and these conditions were first satisfied by Koch in his studies on anthrax. These conditions, often known as 'Koch's postulates', are (1) that the specific organisms must always be present in case of the specific disease, (2) that it must be possible to isolate and culture the organism in a pure state, (3) that a culture of the pure organisms obtained in this way must be able to produce the disease on inoculation into an uninfected animal.

These three criteria were first clearly established in the case of anthrax. One possible criticism, raised by opponents of the germ theory, was that the organisms were obtained in blood from the dead animal, that they were then cultured in a blood-containing medium and this was eventually inoculated into the experimental animal; there was therefore the possibility that blood possessing some unknown toxic components, other than the organism, had been passed from the initial infected animal through cultures to the experimental test animal. This possibility was disproved by Pasteur who sub-cultured drops of blood from an anthrax cadaver in 50 cc. of urine, and then sub-cultured the organism through 100 transfers in urine, giving a final dilution of 100^{100} of the original blood sample.

The development of microbiology

The original drop of blood taken from the anthrax cadaver would thus have been diluted out and yet the final cultures were capable of producing anthrax in the experimental test animal. It proved impossible to produce anthrax in hens by inoculation with cultures of the 'anthrax organism'; a result which needed explanation if the germ theory were to be accepted. Again Pasteur was responsible for seeing the fault in deductions from the experiment: the anthrax organism is unusually heat-sensitive and the hen is an animal with an unusually high temperature, so Pasteur surmised that the temperature of the hen was too high to allow the anthrax organism to proliferate in its tissues and thus cause the disease. He therefore took a hen and immersed it in a bath of cold water; on inoculation the cooled hen developed anthrax and died. Controls were carried out on hens cooled but not inoculated; these survived.

MICROBIOLOGY IN THE TWENTIETH CENTURY

By the beginning of the twentieth century the work described in the previous pages had established that micro-organisms were produced from micro-organisms, that there were many different sorts of micro-organism, and that each particular sort of micro-organism had a specific reaction upon its environment which might appear as a certain chemical change or result in a certain disease in an infected host. Methods for studying the growth of micro-organisms and obtaining them in pure culture had been worked out and it was known that some micro-organisms lead an anaerobic existence while others can live and grow in synthetic media initially devoid of organic matter. Above all, adequate methods of sterilization had been developed so that it was possible, on the one hand, for scientists to work with pure cultures and maintain these cultures in a pure condition and, on the other hand, for doctors and surgeons to work under aseptic conditions so that there was no need to kill organisms after they had invaded the area of operation. Micro-organisms were established as living creatures worthy of study in their own right, forming, as suggested by Haeckel in 1866, a third kingdom of organisms, *Protista*, to rank alongside the existing kingdoms of Plants and Animals. The development of microbiology during the present century has been largely centred around the understanding of living processes as manifest in certain micro-organisms, particularly amongst the bacteria and fungi, which offer experimental advantages in speed of growth, ease of handling and apparent cytological simplicity.

Microbiological research during the first half of the twentieth century was summarized at the inaugural meeting of the Society for General Microbiology in London in 1945 by Stephenson who pioneered the subject of chemical microbiology. She pointed out that five 'levels' of investigation could be distinguished, whether the history of the subject as a whole or the development of any particular problem within the subject was being considered. First came the level A of Initial Exploration in which mixed cultures were studied in natural environments. Then came level B, initiated by Pasteur and Koch, in which the activities and properties of pure cultures in laboratory media were studied. The great development of medical bacteriology took place within this level when the establishment of specificity between organism and disease led to a period of furious activity in which pathogenic organisms were isolated, classified and their relationship to disease studied in detail. Isolation, identification and classification of microorganisms continues actively; today we are not concerned so much with what organism causes which disease, but which particular strain or variant of that organism is causing a particular outbreak of a disease or, in a parallel field, what particular strain of an organism is carrying out a particular chemical reaction.

Investigations in these two levels cannot reveal the mechanisms whereby the effects described are brought about since pure cultures growing in even simple media accomplish a vast variety of different reactions concerned with the utilization of the environment for the production of cell material. The further levels of investigation involve simplification of the biochemical system and alteration of the cell's reactions to its environment. The first simplification was introduced by Stephenson, Quastel and their co-workers in 1920 when they discovered that some bacteria could be centrifuged out of culture, the cells washed in water and suspended as a 'washed suspension of resting cells' which retained many of the activities of the cells in culture. The breakdown of substances by such suspensions could be studied under simple conditions in the absence of cell growth and investigations at this level C gave, for example, information concerning the fermentative and oxidative activities of bacteria and yeasts. The use of poisons and fixatives also enabled some understanding to be gained of the intermediate steps in the overall reactions. Complementary knowledge concerning the pathways utilized for the biosynthesis of cellular material was obtained in the level D of Nutritional Studies in which pure cultures were grown on chemically defined media. In 1890 Winogradski had shown that some organisms, the autotrophes, could grow in a purely inorganic medium whereas others, the heterotrophes, required organic nutrients sometimes of a highly complex nature. Studies of the growth requirements of heterotrophic organisms began with the work of Fildes, Knight and their colleagues from 1930 onwards. They found that some heterotrophic bacteria could grow on a simple medium containing inorganic ions, ammonium as nitrogen source and glucose as source of carbon and energy. Other species were unable to grow on such a medium unless it was supplemented by an amino acid or one of the B group of vitamins. Still others, particularly those found in chemically complex habitats, required several amino acids and several vitamins before they could grow. Fildes and his colleagues postulated that the nutritional requirements were a reflexion of the synthetic disabilities of the organisms concerned: that an organism which required a specific amino acid for growth had, in the course of its evolution, lost the ability to synthesize that amino acid which was nevertheless an essential part of its cellular material. This postulate meant that nutritional studies could be used to study stages in biosynthesis.

Investigations in levels C and D provide indications of metabolic pathways occurring within microbial cells. In level E the biochemist takes over, breaks open the cells, studies the cell-free systems, fractionates and purifies the enzymes involved and so lays bare the mechanisms responsible for the metabolic pathways. This was the latest level defined by Stephenson in 1945. However a further level had already been initiated in 1940 by the studies of Tatum, Beadle and their colleagues, who found that treatment of micro-organisms with X- or ultra-violet radiation increased the rate of mutation to the extent that large numbers of mutants could be isolated and their requirements characterized. Nutritional studies could then be carried out with mutants of the same organism; since each mutant was found to have lost the activity of one enzyme and there was no apparent reason why the range of mutants should not eventually cover all enzymes, it was possible to begin to plot all the steps in biosynthetic pathways and then to work in level E on the properties of the enzymes catalysing each of these steps. Further than this, the fact that mutation resulted in loss of enzymic activity opened the way to an investigation of the nature of the control of enzyme formation and activity. So began the level F of Genetic Studies of unicellular organisms with all the implications that these have had for the advancement of modern biochemistry.

The production of an effective lens by Leeuwenhoek revealed the existence of micro-organisms but the degree of resolution of the best microscope is limited by the wavelength of light. The light microscope can therefore reveal little of the structure of the smaller micro-organisms and nothing at all of the viruses. Fine structure could not be observed until the electron microscope became available and methods were developed for preparing biological material for examination in that instrument. During the last ten to fifteen years great advances have been made in our understanding of the anatomy of cells and it has become clear, amongst other things, that there are two types of microbial cell. There are the relatively simple, procaryotic cells, typified by bacteria and the smaller algae, in which the nucleus is not separated from the cytoplasm by a membrane and does not show differentiation into chromosomes. The larger micro-organisms, such as the protozoa, the fungi and the higher algae, resemble the cells (eucaryotic) of higher organisms in general in that the nucleus differentiates into chromosomes and is separated by a nuclear membrane from a cytoplasm that itself contains organized structure or organelles; the electron microscope has shown that cells contain organelles and particulate material and has, in turn, given impetus to the biochemist to separate such structures so that he can study their functions in relation to the life of the cell as an integrated whole. All levels of experience can now turn to interpretation of living processes in terms of cytological organization and molecular biology as a prelude to what is presumably the next level of investigation: integration and control.

SOME BIOCHEMICAL SIGNPOSTS
IN THE PROGRESS OF NEUROLOGY

by Kendal Dixon

Tell me where is fancy bred,
Or in the heart or in the head?
How begot, how nourished?
Merchant of Venice, III, ii, 64

This vexing problem was scarcely answered by Bassanio when he so successfully selected the leaden casket.

The generation of human thought and the nature of the human mind have given rise to speculation since the dawn of thinking man's existence. It is indeed unlikely that the mind of man, in spite of continuous advances in knowledge, is capable of ever comprehending, even to its own limited satisfaction, more than a small fraction of its internal working. Moreover this imperfect picture of the mind, and of its dwelling house the brain, will always be restricted and distorted by the limitations inherent in human mental mechanism.

However, our knowledge of both brain and mind has expanded remarkably in the last four hundred years. At different points in the history of man the progress of knowledge has been and will be achieved by different kinds of scientific and philosophical discipline—now one and now another. At the present stage the distinction between the architectural and chemical characteristics of living things is rapidly losing significance; the conceptual barrier or demarcation between structure and function, which for so long impeded advance in biology, will soon disappear when life can be interpreted in terms of macromolecular orientation and micromolecular events. In this phase of human progress, therefore, the symbolism which we call biochemistry is specially appropriate and profitable as a technique for the study of cerebral processes.

Some signposts in the progress of neurology

Doubtless when this technique has opened up a more thorough understanding of the molecular statics and dynamics of the brain, other equally intricate problems will be uncovered which may demand entirely new methods of analysis. But at the moment a biochemical approach to this subject appears likely to be particularly fruitful.

How far have biochemical concepts up till now contributed to our comprehension of the mind, its location and its continuity? Some of these signposts in the history of neurology are considered in this lecture.

EARLY CONCEPTS AND THEIR AUTHORS; SITE OF THOUGHT IN THE BRAIN

The ancients believed that Pallas Athene was born from the head of Zeus, thus tacitly giving assent to the concept that Wisdom with her gleaming eyes (γλαυκῶπις ᾿Αθήνη) originates in the head at least of a god. Ideas were necessarily vague until accurate knowledge of the structure of the brain became available, although as early as *c.* 500 B.C. Alcmaeon of Croton regarded the brain as the seat of both sensation and intellect (Guthrie, 1945, p. 49).

The brain controls the body and is fed by the blood

Andrea Vesalius, 1514–64, the great Flemish anatomist and surgeon, was the first to formulate some of the precise ideas in neurology. He was born in Brussels where his father was court pharmacist to the Emperor Charles V. At the age of fifteen, Andrea entered the University of Louvain. He later studied in Paris but then returned to Louvain, where his first memorable dissections of human bodies were carried out. By 1537 his distinction as an anatomist was widely recognised and he was examined by the University of Padua, which awarded him the degree of 'dottore in medicina cum ultima diminutione' and forthwith appointed him professor of surgery and anatomy. Subsequently he resigned his chair at Padua and became court physician to the Emperor Charles V—an early example of a medical man being lured from academic life into the civil service; this step he may have regretted later, since he soon came under Charles' son and successor, the King of Spain, Philip II. In Madrid, at that time in the throes of the Inquisition, it has been said that Vesalius 'could not lay his hand on so much as a dried skull, much less have a chance of making a dissection' (Stirling, 1902, p. 4). But in his early days in Belgium and in Italy he made such important discoveries and deductions from his dissection of human bodies and his experiments on animals as to place him for all time among the most illustrious pioneers of both anatomy and physiology.

His great work *De humani corporis fabrica libri septem* was published in Basel in 1543. It is dedicated to Charles V whom Vesalius regarded with affection and admiration. The seventh book* of this elegantly and lavishly illustrated masterpiece is devoted 'to the brain the seat of mental power and to the organs of the senses'. In this book (Vesalius, pp. 622–3) he suggested that the vital spirit (brought to the brain by the arteries) is adapted for the use of the brain, and from this vital spirit the animal (mental) spirit is elaborated by the activity of the brain. He believed that much of the animal spirit then traverses the ventricles of the brain and is subsequently distributed by the dorsal medulla and thus into the nerves, which arise from this structure. He also supposed that the animal spirit passes through the nerves to the organs of sensation and voluntary movement. In an admirably phrased sentence he made the brilliant speculation that this spirit of activity might either travel *through* hollow passages in the nerves or else might be directed *along* the sides of the nerves in a manner similar to the passage of light in the vault of heaven. He finally concluded that, in any case, the power of the brain reaches the various parts of the body by the continuity of the nerves, as far as he was able to follow by means of surgery on living animals; his experiments indeed showed that the muscles of animals no longer contract at will when the controlling nerves are severed.

Vesalius thus believed that the brain is the master correlator of sensation as well as the initiator of will and movement; he

---

* *Andreae Vesalii Bruxellensis de humani corporis fabrica liber septimus, cerebro animalis facultatis sedi et sensuum organis dedicatus.*

suggested that the brain is nourished by the blood and dispenses its messages to the muscles through the nerves. He showed that, when the nerve to a muscle is cut, the muscle can no longer contract at will although it can still contract when stimulated directly. In fact the dominant influence of brain on the rest of the body was clearly recognised by Vesalius.

By the end of the sixteenth century these ideas of Vesalius had spread widely; by then, informed opinion had reached the conclusion that: (*a*) the seat of will power and thought is in the brain, and (*b*) the brain is nourished by the blood. In fact such concepts had even percolated into the general literature of the age. Thus William Shakespeare said, through the medium of Prince Henry, in a vivid and masterly clinical description of his dying father King John:

> It is too late: the life of all his blood
> Is touch'd corruptibly; and his pure brain,—
> Which some suppose the soul's frail dwelling-house,—
> Doth, by the idle comments that it makes,
> Foretell the ending of mortality.
>
> *King John*, v, vii.

The brain a machine; the body an automaton

René Descartes, 1596–1650, further developed the ideas of Vesalius. Descartes was one of the most outstanding philosophers and mathematicians; he is well known to us all for his brilliant concept of existence being dependent on thought. His contributions to our immediate theme were only of secondary importance; but some of his suggestions about the mechanism of the central nervous system were so ingenious and were expressed with such clarity that we must consider them briefly, in spite of their almost entirely speculative nature.

Descartes subscribed to the views of Vesalius on the nutrition of the brain by the blood and the controlling influence of the brain on the body through the action of the nerves. His first book on the brain *Les passions de l'âme* was written in 1647 for the private use of Princess Élisabeth and was published in 1650. In this brilliant treatise he provided a precise picture of the entry of substances from the blood into the brain; this was one of the earliest descriptions of diffusion. He also gave a graphic account of reflex action. For this he cited the example of eyes which blink when menaced in pretence by a feigned blow. He emphasized that, although we know the potential striker to be a 'friend', who will protect us from any possible harm, all the same we have great difficulty in preventing our eyes from closing. He considered that our bodily mechanisms are so constructed that the movement of a hand towards our eyes excites another movement in our brain so that the ocular muscles involuntarily contract and close our eyelids; he stressed that this movement is automatic and involuntary.

These ideas were expanded in his *Traité de l'homme* published posthumously in Paris in 1664. In this great book he suggested that parts of the blood enter the brain which they feed and maintain, and here they also produce a 'certain kind of very subtle wind, or rather a very active and pure flame which is called the animal spirits'. He believed that the animal spirits are principally produced in the pineal gland in the centre of the brain, from which they enter the cerebral ventricles; the animal spirits then penetrate pores in the cerebral substance, and from thence travel through similar pores in the nerves, which finally conduct the spirits to the muscles. He pictured the movements of the body as controlled in the same manner as are the fountains in the royal gardens; here the force alone, with which water gushes from its head, is sufficient to work various machines and even to make instruments play or speak some words according to the layout of the pipes carrying the water.

He further suggested (in his *Traité de l'homme*) that nerves can well be compared with the pipes in these fountains, the muscles and tendons with the different kinds of mechanism and springs which work these instruments, and the animal spirits with the flowing water itself. He vividly described the effects of external objects on such a human nervous machine; he supposed that each stimulus produces a specific response just as if the visitors who enter his imaginary water garden are forced 'to step on certain tiles so arranged that if, for example, they advance towards a bathing Diana she immediately hides herself in the reeds, and if they try to persue her, they attract towards them a Neptune who threatens them with his trident'. Descartes concluded that when the *thinking mind* is in this machine it will have its principal seat in the brain, and there, like the foreman of the fountains, it can excite, or inhibit, or modify the resulting activity of the body. For Descartes the brain was the central guiding mechanism of man whom he regarded as an automaton.

Descartes' mental picture was superb, but it was an ephemeral castle founded on fancy. His main error was to house imagination, common sense and memory in the pineal gland. But his views on the mechanism of cerebral nutrition, reflex activity, and memory exhale an uncanny aura of modernity. The figment of localizing the highest functions of the brain in the pineal gland was soon opposed by the work of his great contemporary Thomas Willis.

The cerebral cortex

Thomas Willis, 1621–75, the first man to appreciate the importance of the *cortex* of the cerebral hemispheres, was born in Wiltshire. He was educated at Christchurch and took his D.M. at Oxford, where later he became Sedleian Professor of Natural Philosophy.

For most of his active professional life Willis was a busy and fashionable physician in London. He was by no means a toady, however. When called on to see as a patient a young son of James Duke of York (afterwards James II), he gave a gloomy and unflattering opinion about the boy's condition: this he summarized by the words 'mala stamina vitae'. Naturally he was not again consulted by the Court. He was the first to recognise the sweet taste of the urine in diabetes, and he separated into a group those types of diabetes in which the urine is sweet (diabetes mellitus); he was in fact one of the first clinical biochemists. He is known to every medical student to this day for his original description of the arterial supply of the brain (Circle of Willis); he was also the first to give a precise account of the cranial nerves (Nerves of Willis).

In 1664 Willis's most celebrated book *Cerebri anatome cui accessit nervorum descriptio et usus* was published in London; its title page is shown in plate 1. This book is a milestone in the history of neurology. Its translation into English along with translations of Willis's other works appeared in 1681 six years after his death. These translations were bound in one volume entitled *The remaining medical works of that famous and renowned physician Dr. Thomas Willis.* Part VI of this treatise is the translation of *Cerebri anatome*; this section entitled *Of the anatomy of the brain* contains singularly knowledgeable statements some of which we must now consider.

Willis was entirely confident about the site of thought. He wrote 'in truth within the womb of the brain all the conceptions, ideas, forces and powers whatsoever both of the rational and sensitive soul are formed, and having there gotten a species are transformed into acts' (VI, p. 77) and thus subscribed to the views of Vesalius and Descartes.

He next considered the convoluted surface of the cerebral hemispheres:

if it be inquired into, what benefit its turnings and convolutions afford to the brain...the superficies of the brain or cortical substance is uneven and rough with folds and turnings about...For the anfractuous or crankling brain like a plot of ground, planted everywhere with nooks and corners, and dauks and mole-hills, hath a far more ample extension, than if its superficies were plain and even. For as the animal spirits, for the various acts of imagination and memory, ought to be moved within certain or distinct limited or bounded places, and those motions to be often iterated or repeated through the same tracts or paths: for that reason, these manifold convolutions and infoldings of the brain are required for these divers manners of ordinations of the animal spirits, to wit, that in these cells or store-houses, severally placed, might be kept the species of sensible things, and as occasion serves, may be taken from thense. Hence these foldings as rollings about are far more and greater in man than in any other living creature, to wit, for the various and manifold actings of the superior faculties.

Willis thus regarded the gyri of the brain as an effective structure for increasing the areas concerned with thought, imagination and memory, while he considered the action and movement of the animal spirits as the motive power for these activities.

He discriminated (VI, p. 92) between the functions of grey and white matter in a memorable argument:

But that in more perfect animals, all the turnings about are made of a twofold substance, viz. cortical and medullary; the reason seems to be, that one part may serve for the production of the animal spirits, and the other for their exercise and dispensation. For we may well think, that the animal spirits are wholly or for the most part procreated in the cortical substance of the brain; for this severs and receives immediately from the blood the most subtil liquor, and imbuing it with a volatile salt, exalts it into the very pure spirits. It is obvious to everyone, that the arteries enter the cortex of the brain with a more frequent insertion of shoots, and instil to it a spirituous liquor; the leavings of which, or what is superfluous, the veins in like manner entring it, do sup up and carry away; in the meantime, the more subtil portion being here set free, goes into the spirits.

Willis thus for the first time envisaged the dominant action of grey matter in the synthesis of neuronal constituents and was the first to emphasize the importance of the nutrition of the cortex by its singularly copious arterial supply.

His most significant original advance was his idea of the function of the cerebral cortex. Here we must again revert to his own words (vi, p. 96).

Then if the same fluctuation of the spirits is struck against the cortex of the brain, as its utmost banks, it impresses on it the image or character of the sensible object, which when it is afterwards reflected or bent back, raises up the memory of the same thing... And sometimes a certain sensible impression...striking against the cortex of the brain itself...and so induces memory with phantasie.

The original Latin text (of part) of this brilliant theory is illustrated in fig. 1.

Willis's knowledge of anatomy made him discard Descartes' erroneous views. He discounted the idea that the pineal gland is the site of imagination, memory and thought with the words (vi, p. 106) 'animals, which seem to be almost quite destitute of imagination, memory, and other superior powers of the soul, have this glandula or kernel large and fair enough.' He further emphasized (vi, p. 127) that the nerves do not contain cavities like veins or arteries and that the substance of the nerves is 'firm and compacted'; he did however adhere to the Cartesian idea of fluid permeating the nerves.

5-2

Kendall Dixon

Willis has been accused of vagueness and naiveté. Some of his writing is indeed obscure; but this is better than the erroneous precision of Descartes. We should remember that Willis was writing in the middle of the seventeenth century and, although his recognition of the importance of the cerebral cortex was two hundred years in advance of his times, he was

Circa priores advertimus, quòd quoties animæ parte exteriori perculsâ, *Jenfibilis imprefsio*, velut *ſpecies optica*, aut tanquam *aquarum undulatio*, interiùs vergens ad *corpora ſtriata* defertur, *ſenſionis* exteriùs habitæ perceptio five *ſenſus internus* oritur : quòd ſi ifthæc *imprefsio* ulteriùs provecta *corpus calloſum* trajiciat, *ſenſui imaginatio* ſuccedit : dein ſi eadem *ſpirituum fluctuatio Cerebri cortici*, quaſi ripis extimis, allidatur, objecti ſenſibilis *iconem* five *characterem* iſti imprimit, qui cùm exinde poftea reflectitur, ejuſdem rei *memoriam* refuſcitat.

Fig. 1. Reproduction of text from *Cerebri anatome* (1664) in which Thomas Willis states that the image or character of a sensible object raises up memory of the same thing in the cerebral cortex

thoroughly schooled in the medical parlance of his contemporaries. Views like the following (Willis, IV, p. 33) were still current even among the most erudite scholars of this period—
'I knew a certain ingenious man of very hot brain, who affirmed, that after a very plentiful drinking of wine, he was able in the darkest night to read clearly; from hence also may be collected, how the accension of the blood, like the burning of liquors, is increased or made stronger.'

Arguments like this, which occur not infrequently in Willis's works, were bitterly criticized by Michael Foster (1901) in his lectures on the history of physiology. Foster did his best to denigrate Willis and subscribed avidly to the view that the dissections, from which Willis's knowledge was derived, were really carried out by Richard Lower. Be this as it may, it was Willis alone who made intelligent use of these dissections to advance the entirely original suggestion that thought takes place in the cortex of the brain. It may hardly be fair to suggest that Foster (a Cambridge physiologist) was unduly allergic to an anatomist from another university 250 years before; but Foster was at the start of that most unfortunate period of divorce between study of structure and function to which we have referred.

Willis made three suggestions of fundamental importance to the progress of neurology:

(1) the cortex of the brain is the initiator of thought and memory;

(2) the cortex is abundantly fed by the blood;

(3) the cortex dispenses food through the nerves and translates thoughts into action through the nerves.

His main error was to confuse the supply of food to nerves with the passage of nervous messages. His brilliant concept of grey matter as the principal synthetic site in neural metabolism was remarkably substantiated two hundred years later by the work of Augustus Waller. Willis indeed displayed the fortunate foresight of genius.

The cerebral medulla; white matter

Nicolaus Stensen, 1638–86, was born and educated in Copenhagen. He was the most gifted anatomist of his time. He discovered the duct of the parotid gland (Stensen's duct). He also dissected the nervous system with outstanding skill and intuition. Indeed his concept of the structure and function of cerebral white matter still formed the kernel of neurological instruction a hundred years later.

In spite of his great distinction Stensen was not appointed to the chair of anatomy at Copenhagen, when an appropriate vacancy first occurred. He then migrated to Paris, where in 1668 he gave his celebrated *Discours sur l'anatomie du cerveau* at a private meeting in the house of M. Thevenot. This *Discours* was first published only in 1732, when it was incorporated with due acknowledgement into a text book of anatomy by Stensen's great compatriot Jacobus-Benignus Winslow. Winslow had also migrated to Paris, where he was professor of anatomy and surgery. Stensen's discourse forms a chapter in vol. IV of Winslow's treatise; this book saw several editions and was translated into English.

In 1672 Stensen was at last appointed professor of anatomy in Copenhagen; but by then his interests had largely veered to geology, of which science he was indeed one of the prime founders. In 1674 religious zeal brought him to Florence where he became a Catholic priest. He later was made a bishop by the Pope. He died in 1686 worn out by his unflagging missionary activities in northern Europe—one of the greatest scientists of the seventeenth century.

Stensen's discourse on the anatomy of the brain is a model of lucid thinking, originality and modesty. Acknowledged as the leading neuroanatomist of his time he commenced his discourse with the statement that he really knew nothing about cerebral structure, and he remarked that those who imagine they know much about the brain 'speak with the same assurance which they would have if they had penetrated all the designs of its great Architect'. He emphasized the soul's strange impotence: that, having no limits in its external knowledge, it knows nearly nothing about itself. He noted that the diversities on the surface of the brain stimulate wonder 'but when you come to penetrate right inside the brain you see nothing there at all'.

Stensen firmly believed the brain to be the principal 'organe' of the soul; he knew that the brain is composed of two substances, grey and white matter, and considered that the white matter is continuous with the nerves which distribute it throughout the body. He made a profound study of the nature of white matter. He said that to regard the white matter as merely a uniform substance 'like wax, where there is no hidden structure, is to have too low an opinion of the greatest masterpiece of nature'. He suggested that if white matter is everywhere composed of *fibres* 'as it appears certainly in several places so to be, then you must agree with me that this disposition of these fibres must be arranged with a great art, since all the diversity of our thoughts and movements depend on it'. He went on

we admit the intricate structure of the fibres in each muscle, how much more must we admire this intricacy in the brain where the nerves confined to so small a space each operate without confusion and without disorder...It is true that this method is so full of so many difficulties that I do not know if we should ever dare to hope to succeed with it without very special preparations. Its substance (the white matter) is so soft, and the fibres are so delicate that we hardly touch them without breaking them.

The last sentence would be voiced today by every pathologist concerned with examining fresh brain. Fixation of tissues was then unknown.

Stensen's great advance was his discovery of the fibrous nature of the white matter, which he showed to consist of tracts of fibres in continuity with the nerves. He further realized that all thought and movement depend on the arrangement and co-ordination of nerve fibres within the brain. His main error was to oppose the views of Willis, who he said 'lodges memory in the cortex': this was unfortunate, since Stensen's reputation was so great and his integrity so transparent, that the ideas of Willis on the function of the cortex were largely discounted for more than a hundred years. Stensen's criticisms were, it is true, levelled equally at the wild speculations of Descartes and at the more sober assumption of Willis. For Stensen nothing in science was good enough unless proved by rigid argument and meticulous observation. His main contribution was to emphasize the intricate intercommunications of the fine fibres of the white matter as part of the internal mechanism of the thinking mind.

By the latter part of the seventeenth century it was thus generally believed that the brain is the site and centre of thought and that the fibres of white matter are responsible for executing the ordered chains of activity involved in imagination and volition. Owing to continued ignorance of the true nature of grey matter, the views of Willis (that the cerebral cortex is the

seat of memory and the central initiator of mental process)
were not generally accepted. The next hundred and fifty years
brought little further advance to man's knowledge of cerebral
mechanism, although this period was the age of reason, the age
of Locke, Newton, Berkeley and Hume and the age in which
men like Priestley, Lavoisier and Dalton raised chemistry from
alchemy to science. Indeed until more was understood about
the structure, composition and disease of nervous tissue, such
progress was impossible.

INTERMEDIATE ADVANCES IN CEREBRAL ANATOMY AND PATHOLOGY

Although there was little direct progress in knowledge of
intracerebral events during the eighteenth and early nineteenth
centuries, some fundamental advances in neuroanatomy substantially aided appreciation of cerebral function. Thus the
discovery of the separate motor and sensory actions of the
anterior and posterior spinal roots reinforced the idea of the
brain as a co-ordinator of bodily activity. Moreover, accurate
observations on cerebral lesions, in relation to resultant clinical
symptoms and signs, clarified the *normal* function of the parts
injured by disease. Some of these advances will now briefly be
considered; they heralded the accelerated and more spectacular
progress of the second half of the nineteenth century.

Nervous nutrition; motor and sensory nerves

Alexander Monro, 'secundus' (1733–1817), a famous Scottish
anatomist, carried out elegant experiments on the nutrition of
nerves; he also discovered the sensory ganglia on the posterior
roots of the spinal nerves and the interventricular foramina in
the brain (foramina of Monro).

He was the second of three medical men of this name (father,
son and grandson) who successively occupied the chair of
anatomy and surgery at Edinburgh (from 1719 to 1859). His
book *Observations on the structure and functions of the nervous system*
was published in Edinburgh in 1783, and in it describes a new
experimental approach to neurology. He divided the sciatic

nerves of frogs on one side and used the undamaged contralateral nerves as controls. From these experiments he erroneously concluded that there is no change distal to the division and that nutriments therefore do not pass down the nerve. With the microscopes available at the time such conclusions were hardly surprising. Indeed with the optical equipment at his disposal

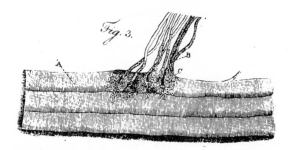

Fig. 2. Illustration by Alexander Monro (1783) of nerve fibres
entering striated muscle

his microscopic observations were truly remarkable: fig. 2 depicts nerve fibres entering a muscle drawn by him and 'viewed with a double microscope, which increases their diameter 146 times'.

Monro had a singularly clear understanding that a nervous impulse does not require the passage of fluid along the nerve; he thus discarded the theory of Descartes. He was also at variance with the views of Willis and others that nourishment passes along nerves. Monro showed that the red dye madder can pass from the blood stream into the growing bones of a young animal, but that the nerves cannot similarly absorb this dye from the blood. From this observation, as well as from his results on the section of nerves, he logically concluded that arteries directly prepare and secrete nourishment and that nerves do not conduct nourishment. It was not for seventy years that this view had to be modified as a result of the brilliant and intricate researches of Augustus Waller—an eventual return to the views of Willis.

Monro's discovery of the ganglia on the posterior roots of the spinal nerves paved the way for his fellow Scot Charles Bell (1774–1842) to show in 1811 that the anterior roots, which do not pass through the ganglia, are alone concerned with *motor* action,* and also for the final proof by François Magendie (1783–1855) of the sensory action of the posterior roots. In fact Sir Charles Bell wrote to his brother in 1826 'two nerves are necessary to a muscle, one to excite action, the other to convey the sense of that action'...one 'carries the will outward'...the other 'conveys inward a sense of the condition...there must be a circle established betwixt the brain and a muscle'. By the early nineteenth century the function of the brain in correlating sensations and voluntary movements through the cerebrospinal nerves was thus fully appreciated.

Disease of the brain

Traumatic injury. Some of the first valid concepts of neuropathology originated with Albrecht von Haller (1708–77), a Swiss nobleman, who was born at Berne. Von Haller studied medicine at Tübingen and later at Leiden, where he took his M.D. in 1727. After further travel and study both on the continent and in England he returned to Berne, and started practice as a physician in addition to teaching anatomy. In 1736 he was appointed to be the first professor of anatomy, surgery and botany in the newly founded university of Göttingen. The invitation to join the staff at Göttingen came from George II of Britain, in his capacity as Elector of Hanover; the Elector displayed admirable concern to appoint the most distinguished scholars in Europe to the chairs in his new university.

Von Haller's famous book (*Elementa physiologicae corporis humani*) was published in Lausanne: the first volume appeared in 1757, while vol. IV, which is devoted to neurology, followed in 1766. Liber x, section 7 of this latter volume is entitled

---

* *Idea of a new anatomy of the brain* (London, 1811), pp. 22, 23. 'On laying bare the roots of the spinal nerves, I found that I could cut across the posterior fasciculus...without convulsing the muscles...but that on touching the anterior fasciculus with the point of the knife, the muscles were immediately convulsed... such were my reasons for concluding that every nerve possessing a double function obtained that by having a double root.'

Phaenomena vivi cerebri; this section is a model of able physiological deduction based on clinical observation. Fig. 3 illustrates a portion of the text of this brilliantly conceived section of the book; von Haller here concluded that

the nerves thus perceive and bring impressions of external objects to the brain. These images are stored in the same brain so as to remain vivid and lucid for fifty or even a hundred years if a man survives to such an age...Various kinds of violence and disease of the brain may destroy the permanence of ideas which persist in the brain of the healthy. In more serious (cerebral) injury memory perishes entirely.

§. XXII. *Porro confirmatur, fenfuum impreffiones in cerebrum deferri.*

Sentiunt ergo nervi, & exterorum corporum impreffiones ad cerebrum devehunt. Eæ in eodem cerebro confervantur, ut omnino quinquaginta & centum annis, fi ad eam ætatem homo fupervixerit, vividæ & limpidæ fuperfunt. Alia erit occafio, qua rectius hæc admirabilis idearum vita dicetur. Rem ipfam tunc fatifiat, per experimenta demonftraffe.

Violentia ergo varia, & morbi cerebri eam idearum perennitatem tollunt, quæ in cerebro fani hominis eft.

Graviori in malo, omnino memoria perit, ex amiffo cerebro (*l*), ejusve organi abfce fu, & corpore callofo erofo (*m*); aliove cerebri vulnere (*n*); ex compreffione a gummate (*o*): ex cerebri aquofa degeneratione (*p*); ex connexione fola (*q*); illifoque capite (*q* *); demum ex fufpenfione fola, durabili vitio (*q†*).

Ex

Fig. 3. Reproduction of text from Vol. IV of *Elementa physiologicae corporis humani* in which Albrecht von Haller (1776) stressed the persistence of memory in human brain for fifty or a hundred years and the destruction of memory by cerebral damage and disease

He also noted that pressure on the brain may give rise to coma and that when this pressure is removed consciousness and also memory may be restored.

Von Haller was one of the most distinguished physiologists of the eighteenth century. He introduced the idea of intrinsic irritability of muscle. In neurology he made the error of regarding the cerebral cortex as merely a mass of blood vessels;

he considered the white matter as the sole nervous centre and originator of thought. But his emphasis on the persistence of memory in the brain and its eradication by cerebral disease was a most important contribution to progress.

Local cerebral ischaemia. The next fundamental advance in ideas about the nutrition of cerebral tissues came from the observations of an eminent Parisian physician, Léon Rostan (1790–1866) who first suggested that local interference in the supply of blood can interrupt cerebral activity.

Rostan's book *Recherches sur la ramollissement du cerveau* was published in Paris in 1823. This book contains masterly accounts of the course of cerebral disease as well as the most discerning correlations between clinical and pathological observations. Rostan suggested that cerebral softening may be caused by defective arterial supply to the brain. He graphically described red infarcts of the brain, which he likened in appearance to the dregs of wine. In considering the cause of softening, he said '.this change in the brain seems often to be a senile degeneration, showing great similarity to the gangrene of old age. As in the latter disease cerebral softening appears to be a disorganisation in which the vessels designed to bring blood and life to the affected organs are ossified...by the process of old age'.

Rostan pointed out that cerebral softening may be confined to the cortex of the brain and that this malady is accompanied by loss of memory, loss of voluntary movement, mental derangement, senile dementa, and finally by complete coma. The fundamental suggestion that disease of the cortex could interfere so profoundly with cerebral function, was a notable return to the views of Willis on the dominant importance of the cortex in all mental activity.

Later the work of Rudolf Virchow (1821–1902) established that thrombosis and embolism are common causes of arterial obstruction with resultant ischaemic injury in many organs including the brain. But to Léon Rostan we owe the initial idea that interference with cerebral nutrition abolishes mental activity and that this interruption can be caused by the local occlusion of the blood vessels which supply the brain.

General cerebral ischaemia. Interruption to the supply of blood to the brain can equally result from failure of the general circulation as well as from local vascular closure.

William Stokes (1804–78), a distinguished Irish physician and Regius Professor of Physic in Trinity College, Dublin, emphasized most clearly that failure of the heart itself can so effectively deprive the brain of fresh blood as to interfere with cerebral activity. His colleague Robert Adams in 1827 published an account of a patient, with a pulse severely slowed by cardiac disease, who suffered from repeated attacks of fainting preceded by transient loss of memory. Later Stokes in his great book *Diseases of the heart and aorta* published in 1854 (pp. 321–3) referred to several patients affected by seizures of this kind, which are now called Stokes–Adams attacks. Stokes realized that these sudden bouts of coma, often preceded or followed by mental confusion, are due to deficiency in the supply of blood to the brain caused by failure of the heart. He gave a vivid account of a patient, who on recovery from each attack remained for about an hour so confused that he was 'unable to recognise his most intimate friends, even his wife was mistaken for his mother'. When such patients died their hearts were found to be grossly damaged from fatty change. Stokes concluded that these attacks of mental confusion, loss of memory and coma follow interference with the supply of blood to the brain resulting from general failure of the circulation. He was also responsible for the extremely important concept that *transient* loss of cerebral function may be a direct sequel to even *temporary* cessation of cerebral nutrition.

The dependence of cerebral activity on oxygen supplied by the circulating blood was finally proved in a most penetrating series of researches by Charles-Edouard Brown-Séquard (1817–78). Most of Brown-Séquard's experimental work was carried out in Paris. He later became professor of physiology and nervous pathology at Harvard, and finally in 1878 returned to Paris where he succeeded Claude Bernard in the chair of experimental medicine at the Collège de France. Brown-Séquard established that cerebral and nervous activity temporarily lost owing to cessation of the circulation could be restored by perfusion with oxygenated blood. He was able in this way to restore co-ordinated facial and ocular movements (apparently under cerebral control) by perfusing the carotid and vertebral arteries of a decapitated dog. In numerous similar experiments he was able to revivify spinal and muscular activity. He concluded that the duration of this power of revival was shortest for brain and was progressively longer respectively for spinal cord, nerves and muscles. He emphasized that brain 'after having completely lost its functions and vital properties, can recover them under the influence of blood charged with oxygen'. He also fully appreciated that of all organs the brain is the most vulnerable to lack of oxygen.

REVELATION OF STRUCTURE AND COMPOSITION
1850–1900

By the mid-nineteenth century the ground was at last prepared. Knowledge of biology and chemistry had progressed to a stage when the methods of these sciences could successfully be applied to the nervous system, while the advances in neuroanatomy and neuropathology, mentioned in the previous section, had finally established that the brain is the dynamic centre of all thought and integrated human activity. Fortunately at this juncture neurology and neurobiochemistry attracted to their study some of the greatest intellects of the age. A period of rapid and spectacular advance followed.

Cellular nature of the nervous system

By the commencement of the Victorian era the cellular structure of plants and animals had become established; this concept had been achieved gradually by a series of observations over the preceding two centuries which were finally welded into a composite theory by Theodor Schwann (1810–82). Schwann's work was communicated to the Académie des Sciences in Paris in 1838 and was first published in 1839; an excellent English translation by Henry Smith F.R.C.S. appeared in 1847. In this book Schwann described ganglion-globules in the grey matter of the brain and spinal cord which contain nuclei and nucleoli (pp. 152, 153). These cell-bodies were observed a few years earlier by Remak and by Valentin (see Schwann p. 152); but the general cellular picture of both grey and white matter was first emphasized by Schwann, who also first described the doubled-contoured white substance which forms the sheath of white fibres. Schwann even realized that the material in the sheathes of white fibres is fat-like (p. 148) in nature and is responsible for the white colour of the fibres 'when examined with the unaided eye'. Later work by Purkinje, by Rosenthal and by von Kölliker showed that the fibres are processes emanating from the cell-bodies. In 1867 von Kölliker described the myelin sheath in fresh nerve fibres as 'perfectly homogeneous and viscous, like a thick oil' (p. 239); he considered the glistening appearance of nerves and white matter as dependent on the properties of this material in the nerve sheath. By the time Augustus Waller commenced his researches (see below) the cellular nature of the nervous system was thus at least partly revealed.

The neurone as a metabolic unit

Augustus Volnay Waller (1816–70), was born in Faversham in Kent. Owing to his father's connexion with the wine trade he spent much of his youth in France. This enabled him to write with bilingual ease. His publications in French were indeed rapidly heralded and acknowledged throughout the scientific world abroad, where Waller's work always received greater recognition than among his own compatriots. Physiologists in this country have tended to disregard his fundamental contributions to our knowledge of neuronal metabolism and scarcely ever refer to his great principle of neuronal integrity which the French call 'la loi de Waller'.

Waller qualified Docteur en Médecine of Paris in 1840, and a year later became a Licentiate of the Society of Apothecaries of London. Then for ten years he worked as a general practitioner in Kensington. Thereafter he mainly lived abroad, although for a short period he was Professor of Physiology at Queen's College, Birmingham, and he also held a hospital appointment in that city as physician. He became F.R.S. in

1851. By this time he had already discovered the emigration of polymorphonuclear leucocytes in inflammation and also the vasaconstrictor action of the sympathetic nervous system, in addition to opening up an entirely new vista in his great concept of the cellular integrity of the neurone. Any of these discoveries alone would have justly placed him for always among the pioneers of medical science.

In 1850 Waller's famous paper, entitled 'Experiments on the section of the glossopharyngeal and hypoglossal nerves of the frog and observations of the alterations produced thereby in the structure of their primitive fibres', appeared in the *Philosophical Transactions* of the Royal Society of London. In this article he described the elegant technique which he had devised for the study of nervous degeneration: he cut either the glossopharyngeal or the hypoglossal nerve on one side and later compared the distal ramifications of the severed trunk in one side of the tongue with filaments from the undamaged contralateral nerves. He thus contrasted the degenerating nerves with unaffected nerves from the opposite side of the same animal— an ideal system of controlled experiment. He established that, when 'continuity with the brain becomes interrupted by section', the myelin in the severed *distal* portion of the nerve becomes turbid and coagulated in three-four days and 'curdles into separate particles' within five-six days. Moreover, he found that the healthy nerve tubes swell, when they are immersed in water, but, that the nerve tubes in the distal portion of the severed nerve have lost this power of swelling when placed in water. This was probably the first account of loss of semipermeability in injured or dying cells.

Waller's illustrations of degenerating nerve fibres are remarkably accurate. Plate 2 shows a degenerating papillary nerve in a frog's tongue six days after section of the glossopharyngeal nerve. At this stage the fibres are replaced by contiguous ovoid globules which consist of a double-contoured shell of myelin enclosing a central space. Plate 3 illustrates a later phase in the degeneration of a similar papillary nerve twenty-two days after section: the ovoid globules have now disintegrated into rows of amorphous granules. Waller thus

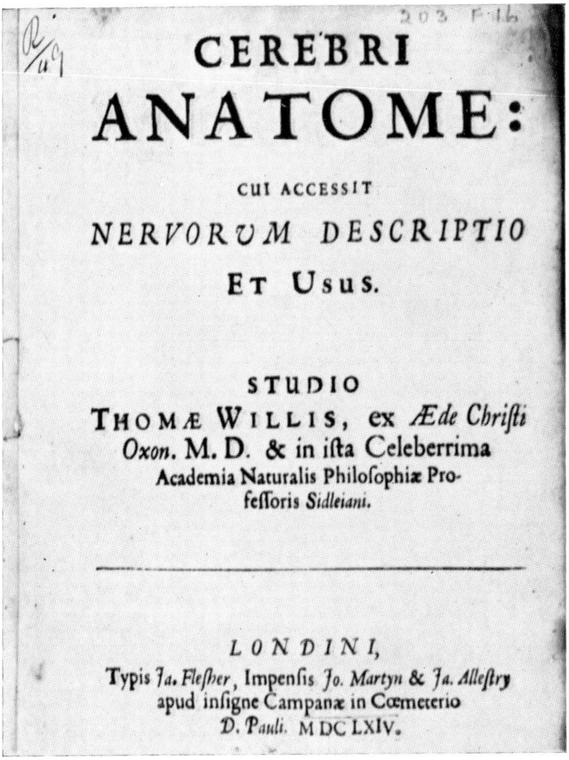

CEREBRI
ANATOME:

CUI ACCESSIT

NERVORUM DESCRIPTIO
ET USUS.

STUDIO
THOMÆ WILLIS, ex Æde Christi
Oxon. M. D. & in ista Celeberrima
Academia Naturalis Philosophiæ Pro-
fessoris *Sidleiani.*

LONDINI,
Typis *Ja. Flesher,* Impensis *Jo. Martyn* & *Ja. Allestry*
apud insigne Campanæ in Cœmeterio
D. Pauli. M DC LXIV.

Plate 1. Title page of *Cerebri anatome* by Thomas Willis

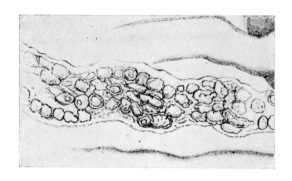

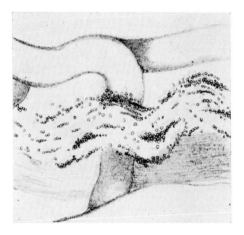

Plates 2 and 3. Illustrations by Augustus Waller (1850) of papillary nerves
in frog's tongue following section of glossopharyngeal nerve

Plate 2. Six days after section: myelin
has disintegrated into ovoid globules,
many of which show a double contour

Plate 3. Twenty-two days after section:
myelin has disintegrated into rows of
amorphous granules

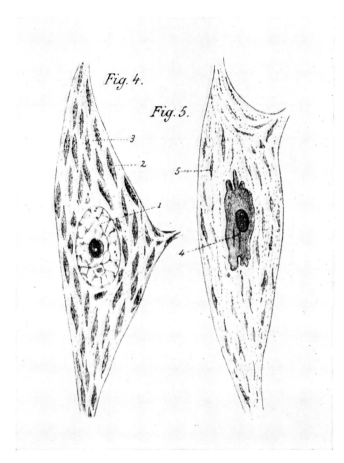

Plate 4. Illustration by Gustav Mann (1895) of Nissl spindles in spinal anterior horn cells: *on left* a cell at rest, and *on right* a cell after excessive muscular exercise

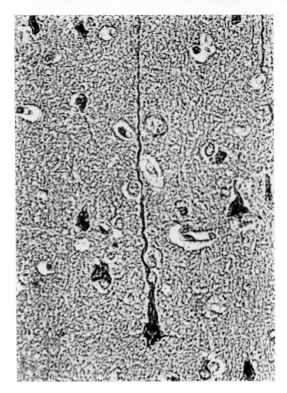

Plate 5. Rabbit cerebral cortex stained by Danielli's tetrazo method for protein. The dendritic branches (in the neuropil) and also a chromophilic pyramidal cell and its apical dendrite give an intense reaction. × 420

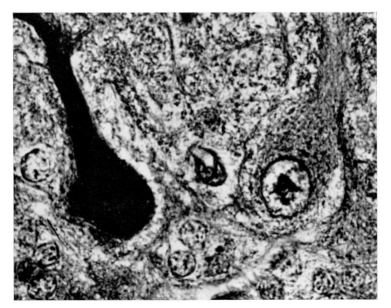

Plate 6. Rabbit cerebellar cortex. Treated with tannic acid (4 per cent. solution in 0·1N HCl) and then with ferric chloride ($FeCl_3.6H_2O$, 10 g./100 ml. water). Tannophilic protein coloured black: *on left* a chromophilic Purkinje cell, *on right* a chromophobic Purkinje cell. × 1060

found that both axis cylinder and myelin sheath completely disintegrate when their connexion with the spinal cord is severed. His description of the process of degeneration accords perfectly with observation made by use of modern cytochemical techniques (Noback and Montagna, 1952).

Waller's next step (1852*a*) was to confirm the earlier discovery of Fontana that a cut nerve can later regenerate. But Waller further demonstrated that all growth of regenerating fibres originates proximally to the point of section and not distally to it. In this concept Waller 'was able to formulate the true solution of the problem, orienting himself without difficulty in a puzzling field in which even distinguished modern scientists have lost their way' (Ramon y Cajal, 1928, p. 8). Waller concluded that continuity of the fibres above the point of section with their cell bodies secures survival and enables the proximal ends of the cut fibres to sprout out fresh extensions along the original course of the severed nerve.

He then performed a remarkable series of experiments (1852*b*) on the transection of nerve fibres so as to break their continuity with the grey matter to which they are normally attached. For this purpose he cut anterior and posterior roots of spinal nerves, the trunk of the vagus nerve in the neck, and also the spinal cord itself. The posterior roots were cut either above or below the spinal ganglia, the section of the vagus nerve was made between its two ganglionic bodies in the neck or on either side of them. After section he studied the resulting structural degeneration in the nerves and in the ganglia as well as any loss in the power of the nerves to conduct impulses.

From these observations he established that all nerve fibres are filiform cellular processes connected with cell bodies (ganglionic corpuscles) and that the fibres degenerate if separated by section from their cell bodies, but that, even when they are cut, the fibres survive if still connected with surviving cell bodies. He suggested that fibres which arise from the spinal ganglia and from the ganglia on the vagus nerve have their *vital nutritive centre* in the cell-bodies situated in the related ganglia. He further discovered that the behaviour of the fibres

6   81

in spinal roots and tracts after section varies according to the differing situation of their controlling neuronal cell bodies. Thus twenty days after section of the spinal cord, together with some of its anterior and posterior roots, he found the posterior white columns of the cord *above* the point of section were disorganized, since the connexion of the fibres with their cell-bodies situated in the posterior root ganglia was severed; but the posterior white columns *below* the point of section remained intact since their connexion with the posterior root ganglia was preserved. Similarly, among the severed nerve roots, the cut anterior roots were disorganized and granular (and no longer conducted stimuli), as their fibres were disconnected from their controlling cell-bodies in the spinal anterior horns, while the posterior roots, although no longer connected with the cord, were intact, as their fibres still were joined to their cell-bodies situated in the posterior root ganglia.

These results led to Waller's brilliant concept of the *nutritive influence* of the cell bodies on the whole fibre (1852*c*)—'le centre nutritif des racines antérieures se trouve dans la moelle épinière, tandis que celui des racines sensitives est dans les ganglions intervertébraux...il faut admettre que les fibres motrices sont en rapport avec des corpuscles nerveux situés...à la moelle épinière'. He thus concluded that the nutritive cell bodies for the sensory fibres are in the spinal ganglia, while, for the motor fibres, the controlling nutritive cells are in the anterior horns (Waller, 1857) of the spinal cord itself. The nutritive influence of the cell body on the whole fibre was thus revealed, and it was at last understood that neuronal processes several feet in length are continuously dependent on the passage of essential nutriment from their cell-bodies.

Waller was also responsible for yet another fundamental idea in biochemistry, namely the equilibrium between anabolism and katabolism. This followed from his work on the rate of degeneration of transected nerves of frogs kept at different temperatures (1852*d*). He found that degeneration and loss of function are retarded in the transected nerves of animals kept at low temperature; he considered that this retardation of the process of degeneration depends on the slower rate of chemical reaction at lower temperature. He had indeed a vivid picture of the equilibrium between anabolism and katabolism; this view is expressed in a memorable sentence 'le corps de l'animal se compose de parties qui se détruisent et se renouvellent sans cesse...cela provient de l'équilibre qui existe entre ces deux actions contraires'. In the transected nerve he believed that anabolism is abolished but that katabolism still proceeds, he suggested that a fall in temperature can retard katabolism and so delay disintegration. He thus pictured a perennial intraneuronal equilibrium between anabolism and katabolism. He postulated the necessity for continuous renewal of neuronal substance depending on the activity of the cell-body. So long as the influence of the ganglion-cell on the fibre persists this equilibrium is maintained; but once the continuity with the cell-body is broken, renewal immediately ceases. Without this renewal the fibre dies, but a fall in temperature, by lowering the rate of katabolism, can retard the onset of death and degeneration. The essential nature of grey matter, as the domicile for cell-bodies and as the cellular co-ordinator of the whole nervous system, is inherent in Waller's discoveries. The views of Willis on the function of grey matter were indeed substantiated at last on a solid experimental basis.

At this epoch in the history of science Waller's discernment was fantastic in its precision.* The dominant synthetic activity of the cell-body and the basic cellular significance of grey matter were thus demonstrated with lucid clarity by this truly brilliant Victorian thinker.

The chemical composition of the brain

The establishment of the atomic theory and the discovery of the molecular composition of matter immediately attracted the attention of biologists and medical men. These new concepts were applied widely and extracts made from brain as well as from other tissues were soon subjected to systematic investigation by chemical methods.

---

* Indeed Ramon y Cajal, writing in 1928, referred to the conceptions of this 'English scientist' (Waller), which were contrary to the later views of some German histologists, with the remark 'error is modern while truth is ancient'. *Degeneration and regeneration of the nervous system* (London), vol. 1, p. 5.

Kendal Dixon

A. F. Fourcroy (1793) was one of the first to study cerebral chemistry. He extracted brain with water and found that the resultant solution coagulates on heating in the same way as egg albumen. He also observed that an aqueous emulsion of brain is coagulated and turned yellow by nitric acid. He further isolated a fat-like substance from alcoholic extracts of brain; he regarded this material as akin to 'a thick oil', which is peculiar to brain and can form needle-like crystals and plaques. Fourcroy thus discovered both the proteins and the lipids of brain.

L. N. Vauquelin, in a masterly paper published in 1812, confirmed Fourcroy's results and showed that the substance extracted from brain by alcohol contains phosphorus. He said 'we must therefore necessarily admit the existence of phosphorus in the substance of the brain'. He concluded that this phosphorus is combined with fatty material; he thus was the discoverer of the phospholipids. Vauquelin also noted that the medulla oblongata and the spinal marrow contain more fatty material but less protein than grey matter.

Further advances by the French school of chemists followed. J. P. Couerbe in 1834 recognized cholesterol as a normal constituent of brain, while Edmond Frémy in 1841 established that white matter is richer in lipids than grey matter. The work of Gobley further demonstrated the similarity between the phospholipids of brain and of egg yolk as well as the presence in them both of glycerophosphoric acid. Lecithin was finally isolated from brain by A. A. Strecker in 1868.

Unfortunately in 1864 Oskar Liebreich, a pupil of Hoppe-Seyler, claimed that brain is composed almost entirely of a single substance called 'protagon', which was supposed to contain nitrogen and phosphorus as well as carbon, hydrogen and oxygen; all other substances isolated from brain, Liebreich asserted, are products of the decomposition of this unique archetypic compound. The myth of protagon was fiercely upheld by Hoppe-Seyler in Germany and also secured powerful adherents in England, namely Arthur Gamgee and his collaborators (see below). Indeed this fantasy was supported by many of the established physiologists of the time including those of

Michael Foster's school such as Sheridan Lea (1892). At this point we must turn to the man who dared to shatter this myth, and who placed solid science in its stead—the greatest of all cerebral chemists—namely Ludwig Thudichum.

Ludwig Johannes Wilhelm Thudichum (1829–1901) was born in the quiet Hessian town of Budingen in 1829.* His father Georg Thudichum was the town's leading Lutheran minister and also principal (Oberstudienrat) of its high school or gymnasium. Georg Thudichum was a Greek scholar of eminence, noted for his translations of Sophocles; he gave his son a good classical education, which is evident in the precise and elegant language of Ludwig's numerous published works.

Ludwig showed early promise of intellectual brilliance; he was outstanding in his medical studies and he also was a highly talented singer and pianist. His medical training was at Giessen and later at Heidelberg. Among his teachers were Liebig, Gmelin and Henle. Liebig indeed was a close friend of the Thudichum family and it was probably his influence which kindled Ludwig's flair for medical chemistry. As a medical student at Heidelberg Ludwig won a prize for a paper on urea in amnionic fluid. He took his M.D. at Giessen in 1851. During the war of 1850–1 against Denmark he served as a volunteer surgeon in the army of Schleswig–Holstein. His overt sympathy, while a student, for a current revolutionary movement, impeded his appointment to civil medical positions in Germany. Disappointed by this discrimination against him, he migrated in 1854 to London, where his fiancée and distant cousin Charlotte Dupré had moved from Frankfurt-an-Main a few years earlier. After their marriage Ludwig and Charlotte made their permanent home in London, being attracted by the liberal atmosphere of nineteenth-century Britain. Thereafter Ludwig Thudichum lived the life of a cultured and respected Victorian doctor.

Thudichum's interest in all branches of medical chemistry was continuous, but so also was his attraction to clinical medi-

* For an excellent account of Thudichum's life and the controversy on protagon the reader is referred to D. L. Drabkin, *Thudichum: chemist of the brain* (Philadelphia, 1958).

cine itself; in fact he remained a practising physician and surgeon all his life. The catholicity of his learning and of his
original work was quite exceptional even in an age of polymaths.
His clinical appointments included the posts of Physician to the
St Pancras' Dispensary and of Visiting Surgeon to the Queen's
Jubilee Hospital at Brampton. As a rhinologist he was in the
first rank; he invented the nasal speculum and introduced
electro-cautery into nasal surgery. He became M.R.C.P. in
1860 and was elected F.R.C.P. in 1878. Simultaneously he was
drawn to the problems of medical chemistry, in which subject
he also held several appointments. He was professor of chemistry
at the Grosvenor Place School of Medicine and subsequently
became lecturer and first director of the Laboratory for
Chemistry and Pathology at St Thomas's Hospital; later he
resigned from his position at St Thomas's so that he could give
more of his time to clinical practice and in this way finance his
researches.

In 1863 Thudichum's great treatise on gallstones was
published. In this book his 'nidus' theory of the generation of
biliary calculi was advanced. His view of the concentric and
laminar formation of gallstones is held to this day. In 1864 he
won the Sir Charles Hastings gold medal of the British Medical
Association for the discovery and isolation of *urochrome*, the
principal colouring matter of normal urine. In 1867 he first
prepared the iron-free pigment haematoporphyrin by the
action of sulphuric acid on haemoglobin. In 1869 he first
isolated and characterized the luteine pigments, now known
as the carotenoids. Even one of these discoveries alone more
than merited his election to the Royal Society of London. But
Thudichum lacked the influence required to secure recognition by the cliques of established scientists in England
and in Germany, and moreover, as we shall see later, vested
interests in these circles soon became actively opposed to
giving his great advances even the smallest modicum of due
acknowledgement.

These earlier discoveries, any of which by themselves would
have placed Thudichum among the great pioneers of medical
science, are indeed outside the great corpus of his work. We

now come to his major contribution—*the chemistry of the brain*. At the time when Thudichum started to work in London, Sir John Simon was Medical Officer to the Privy Council. Sir John's position corresponded to that of Secretary to the Medical Research Council today. Sir John was a man of great vision; he believed that typhus and other diseases affecting the brain might have a chemical explanation. Sir John was so impressed by Thudichum's earlier work that he persuaded the government to support Thudichum's researches on the brain. For this purpose Thudichum was appointed Chemist to the Local Government Board in 1864. The results of his work on the brain were published (1874–82) in the Reports of the Medical Officer to the Privy Council. This remarkable series of articles culminated in the appearance in 1884 of Thudichum's monumental *Treatise on the chemical constitution of the brain*.

This work on cerebral chemistry was one of the most fruitful sequences of research ever conducted by a single individual. First of all he succeeded in isolating kephalin and distinguishing it from lecithin. Sphingomyelin was then isolated and characterized; in fact to him we owe the recognition of the phospholipids as a whole class of compounds of singular biological importance; he introduced the word phosphatide to describe them—'phosphatides' he said (1884, p. xii) 'are the centre, life and chemical soul of all bioplasm whatsoever, that of plants as well as animals'. He realized that these substances are specially abundant in brain. He next discovered, in the brain, an entirely new group of cerebral substances, the cerebrosides; among these he isolated both phrenosin and kerasin. Not only did he purify and analyse these parent substances, but also their constituent parts. He showed that phrenosin is composed of one molecule of a newly discovered base sphingosine condensed with a hexose and with a fatty acid. This hexose was called by him cerebrose; it is now known to be identical with galactose derived from lactose by hydrolysis. He also discovered that the lactic acid in brain occurs mainly in the grey matter, and that it is optically active, and is in fact the same substance as the sarcolactic acid of muscle.

This impressive and indeed unique catalogue of discoveries is summarized in his treatise on the chemical constitution of the brain. He realized immediately the likely bearing of his findings on cerebral function in health and disease. He emphasized (1884, pp. 39, 40), as we have seen, the remarkable properties of the phosphatides—this great class of lipids which he had himself defined. In water they 'form a turbid solution of very peculiar appearance'. Their particles in aqueous solution are mostly 'so small as to be beyond the reach of optical definition as single particles'. Yet they 'exhibit their presence by iridescence...and by reflecting polarised light'. Unlike emulsions of fats with water they require no emulsifying agent, but 'these brain substances give this peculiar solution with mere water'. Thudichum regarded this emulsion 'as a state of imperfect or incomplete solution, a stage intermediate between the solid and the fluid state of matter'. He stressed the peculiar attraction of these particles of phospholipid to water, and their consequent repulsion from each other: he was indeed very close to the conception of micellar structure now known to be such an important characteristic of this class of colloid. He suggested (1884, p. xiii) that some of these phosphatides may be in true solution at body temperature but 'become colloid at temperatures between the normal and highest fever heat. It is not impossible that some such change may be the cause of death in many febrile conditions and in many cases of exposure to excessive heat'. He was thus the first to suggest that thermal injury to protoplasm may be determined by changes in cellular lipids, a view which has received considerable recent support. He even supposed (1884, p. xii) that 'so-called softening of the brain consists in the first place in the loss of the colloid state' and (p. 259) that 'the great diseases of the brain and spine, such as general paralysis, acute and chronic mania, melancholy and others, will be shown to be connected with specific chemical changes in the neuroplasm'. He realized that in tabes dorsalis the cerebrosides of the white matter in the posterior columns of the spinal cord are decomposed, and he made the stimulating suggestion that amyloid bodies of the brain and cord (which are rich in carbohydrate) are formed from the sugar liberated by decomposition of cerebrosides. Thudichum was indeed a chemist of outstanding distinction, who was able from a deep knowledge of medicine to see the full implications of his chemical discoveries on the nature of the brain and its disease.

How was it that these discoveries, which illuminated afresh the whole subject of cerebral chemistry, were almost totally neglected until some so-called leaders of physiological chemistry of Thudichum's age had passed into well merited oblivion? Thudichum was himself neither the pupil nor the assistant of anyone in a key position in physiological chemistry in Victorian Britain. Moreover he had the misfortune to offend the contemporary scientific establishment *both* in Germany and in Britain. His work immediately revealed the error of Liebreich's tenet that a mythical substance protagon is the principal constituent of the brain. Unfortunately this view had powerful adherents. Among these were Felix Hoppe-Seyler, editor and founder of the *Zeitschrift für physiologische Chemie* and Arthur Gamgee, Brackenbury Professor of Physiology at Owens College, Manchester. Hoppe-Seyler doubtless resented the refutation of work of his own pupil Liebreich. Also Hoppe-Seyler may have been jealous of Thudichum's priority in the isolation of haematoporphyrin; this was never properly acknowledged by Hoppe-Seyler. Gamgee bitterly attacked Thudichum's publications and described work which purported to confirm the existence of protagon. Unfortunately, in loyal defence of Thudichum, two of his collaborators made an adverse comment on a current text-book, and they thus offended its authors who were important physiologists of the time. Thus, to use Drabkin's words 'new and powerful enemies were made' (1958, p. 125). Protagon was from then on supported by the establishment until the end of the century. It is particularly humiliating to Cambridge men to realize that the idea of protagon was promulgated, without the slightest reference to Thudichum's brilliant work on the brain, in the apparently authoritative biochemical appendix to *Foster's text-book of physiology* (Lea, 1892).

Although a highly critical commentator on the notion of protagon and justifiably caustic* on the erroneous views of

---

* Thudichum (1883), p. 526. 'It is surprising how little chemical relations of the brain are understood by physiologists and chemists of profession. They ignore

cerebral chemistry held by so many academic physiologists of the time, Thudichum was remarkably unaffected by lack of recognition and by cruel injustice, which would have stunned many a lesser spirit. He knew he was right and he remained one of the most versatile and radiant scientists of his century.

In spite of this grossly unjust denigration by established physiologists of his time, Thudichum was regarded with affection and esteem by his numerous colleagues among the medical profession of London. In this circle his genius, his culture, and the warmth of his character and personality were highly prized; he was the dynamic and stimulating President of the West London Medico-Chirurgical Society. His musical and literary background and, more important still, the affection of his own family also gave him much needed support and calm for his prolonged conflict with the obstinate intransigence of established ignorance.

Thudichum turned his attention to many subjects outside medicine and cerebral biochemistry; all of these he clarified by his penetrating mind and lucid style of writing. This is well exemplified by his *Treatise on wines* published in 1897. This book is illustrated by elegant engravings; in it, in succinct chemical terms, he dissipated much of the fog and confusion concerning the maturation of wines and he also gave illuminating views on the nature of 'bouquet'.

The remarkable quality of Thudichum's work was, that it was based on intricate purifications and analyses, and yet he never lost sight of its broad scientific and medical import. As he said 'the brain is the most marvellous chemical laboratory in the animal economy; in it the albuminous, phosphorised and nitrogenous principles...are brought into the most varied relations for the production of power of the most refined nature' (1883, p. 526). His vision foresaw that various forms of insanity may have a chemical basis. He even suggested (see fig. 4) that insanity may consist of the 'external manifestations of the

the broadcast facts and maintain the most absurd fallacy, namely the so-called doctrine of protagon. They thereby impede the progress of science, and comprise the minds of those who are desirous to learn and to work...Every protagonist has hitherto promised to teach us its rational constitution...There was not only much cry and little wool, but no wool at all'.

effects on the brain-substance of poisons fermented in the body, just as the mental aberrations accompanying chronic alcoholic intoxication are the accumulated effects of a relatively simple poison fermented out of the body' (1884, p. xiii). He

A few further examples may suffice to indicate some of the lines on which the practical consideration of diseases of the brain by the aid of its chemistry will, at least in the first instance, have to proceed. Many kinds of headache are probably due to intra-cranially brewed chemical poisons, or to poisons carried from the body to the brain by the blood, whether fermented in the body, or like alcohol, morphia, and fusel oil, formed out of the body. From such occasionally produced effects to the constant produc-tion of similar effects by a continued zymosis, be it now caused by organised or unorganised ferments, is not a great step. Many forms of insanity are unquestionably the external manifestations of the effects upon the brain-substance of poisons fermented within the body, just as the mental aberrations accompanying chronic alcoholic intoxication are the accumulated effects of a relatively simple poison fermented out of the body. These poisons we shall, I have no doubt, be able to isolate after we know the normal chemistry to its uttermost detail. And then will come in their turn the crowning discoveries to which all our efforts must ultimately be directed, namely, the discoveries of the antidotes to the poisons, and to the fermenting causes and processes which produce them.

THE AUTHOR.

11, PEMBROKE GARDENS, W.,
February, 1884.

Fig. 4. Reproduction of text from preface to Ludwig Thudichum's *A treatise on the chemical constitution of the brain* (1883). Here Thudichum suggested that poisonous substances formed by abnormal metabolism are responsible for mental disease.

was thus the originator of the concepts of biochemical lesions and auto-intoxications so prevalent in psychological medicine today. He actually supposed that such abnormal products may be formed by 'continued zymosis, be it now caused by organized or unorganized ferments'—in fact a deflection of enzymic action. This brief account of Thudichum and his work may appropriately close with the final sentence of his historic book 'In short it is probable that by the aid of chemistry many derangements of the brain and mind, which are at present obscure, will become accurately definable and amenable to precise treatment, and what is now an object of anxious empiricism will become one for the proud exercise of exact science' (1884, p. 259–60).

Cerebral neurones as independent units

The idea of the neurone as the unit of both structure and function in the central nervous system did not become firmly established until the end of the nineteenth century. Waller's brilliant concept of functional and metabolic entity was indeed widely accepted in regard to neuronal nutrition, but this idea was not fully applied either to the structure or the activity of cerebral neurones. Waller's experiments mainly concerned neurones of the spinal cord and sensory ganglia and his results could not easily be translated at the time into terms applicable to the visibly much more complex structure of the higher centres. In fact for long the theory of Gerlach prevailed; this supposed that there exists a network of continuity between the processes emanating from different cell bodies. Grey matter was thus regarded as a complex syncytial reticulum composed of the intercommunicating processes of a multitude of cells.

The separate entity and integrity of the neurone, as both the structural and functional unit of the whole central nervous system, was eventually established by the work of His and of Forel, culminating in the masterly researches of the great Spanish neurohistologist Santiago Ramon y Cajal (1852–1934).

Ramon's work is admirably summarized in his book *Les nouvelles idées sur la structure du système nerveux chez l'homme et chez les vertébrés* which was published in Paris in 1894. His first results were obtained by using the silver carbonate method of Camillo Golgi. Later, Ramon introduced his own now famous silver impregnation—reduction or 'photographic' method for demonstrating the finest neuronal terminals. By use of the Golgi technique, however, he first clearly demonstrated the varicose axonal terminals at the axo-dendritic and the axo-somatic approximations. These neuro-dendritic and neuro-somatic contacts he observed in the grey matter of spinal cord, cerebrum and cerebellum. He concluded (p. 9):

(1) Nervous cells are independent units, they never anastomose either through their dendritic branches or through nerve fibres emanating from their axons.
(2) Every axis cylinder terminates freely in varicose and flexuous arborisations...
(3) These arborisations are applied either to the body or to the dendritic branches of other nervous cells establishing connexion by contiguity...which is just as efficacious in transmitting impulses as if there were real connexion of substance between the neurones.
(4) The cell body and the dendritic branches are as much concerned with conduction of impulses as with neuronal nutrition. The dendrites carry impulses to the cell body, while axonal transmission is away from the cell body.

Ramon established that the *neurone* (a term introduced by Waldeyer) is the 'veritable' cellular unit (p. 171) of the nervous system. He found that the axonal terminals of every neurone end freely, and that these terminals are *never continuous* but *contiguous* with the cell bodies and dendritic branches of other neurones. He denied continuity between cells and insisted that the passage of nervous impulses from the processes of one cell to those of another depend on contiguity or contact but not on continuity of substance.

Ramon accomplished intricate and meticulous analyses of the interneuronal connexions in both spinal and cerebral grey matter. In the cerebellar cortex he discovered the unique axo-dendritic relations between the axonal terminals of the granular cells and the dendritic ramifications of the Purkinje cells; here the axonal terminals traverse the dendrites at right angles so as to penetrate successive arrays of aligned dendritic branches and thus run the whole length of the cerebellar folia so as to make a recurrent sequence of interneuronal contacts.

Ramon's most momentous contribution was to establish similar axo-dendritic contiguity in the cerebral cortex itself. In this tissue Camillo Golgi, indeed, first described the plumate branches of the apical dendrites of the pyramidal cells; but Golgi regarded these dendritic or protoplasmic branches as attachments to blood vessels and as mere nutritive appendages for the neurone as a whole. Ramon, in a brilliant investigation of cortical architecture showed that the apical dendrite of each pyramidal cell, on reaching the molecular layer, subdivides (p. 50) into a 'superbe panache ou bouquet de ramuscules protoplasmiques, se terminant librement entre les fibrilles nerveuses de cette zone'. He designated (p. 52) the pyramidal cell as 'cellule psychique'. He thus established that the terminal plumate dendritic branches of the pyramidal cells, arrayed with the equally complex contiguous ramifications of axonal terminals, together form the dominant cellular components of the thinking mind.

The contents of the neurone: neuronal cytochemistry

Nissl substance. By the end of the nineteenth century the contents of neurones as well as their interconnexions were being actively investigated. Chromophilic granules, which are coloured more intensely by stains than the rest of the cytoplasm, were described by Rudolf Arndt in 1874; but it was Franz Nissl in 1890, who first showed that these characteristic chromophilic regions of neuronal cytoplasm are specifically coloured by basic dyes. The material which is stained in this way is generally called the Nissl, chromophilic or tigroid substance; this last appellation is given to denote the striate arrangement of the ovoid or spindle-shaped particles of this substance observed in fixed cells. Nissl substance occurs in the cell bodies and dendrites, but is absent from the axon; it is stained intensely by the basic dyes methylene blue, toluidin blue and pyronin.

Nissl degeneration. In 1892 Nissl tore out the facial nerve on one side in rabbits. He discovered that, following this injury, the cell bodies of the neurones in the facial nucleus in the pons become swollen and their chromophilic granules disappear. This change is called Nissl degeneration; it involves both neuronal swelling and dissolution of Nissl substance, the latter event being called chromatolysis. Nissl degeneration occurs *proximal* to the point of division or injury of a nerve fibre and is complementary to Wallerian degeneration which takes place in the *distal* segment of the severed axon. Unlike Wallerian degeneration Nissl degeneration is usually a reversible process; in fact many of the injured cell bodies recover, the chromophilic substance is restored, and later (as shown by Ramon y Cajal) axonal sprouts grow from the proximal segment of the divided axon. The restoration doubtless involves intense synthetic activity in the cell body and dendrites and affords striking confirmation of Waller's view that the cell body is responsible for the synthetic or anabolic processes of neuronal metabolism.

Chemical nature of Nissl substance. The basophilia of the Nissl substance indicates that it contains acidic groups. Hans Held in 1895 made observations of fundamental importance on the chemical nature of the Nissl substance. He discovered that the tigroid material is dissolved by sodium hydroxide, but, unlike the remainder of neuronal cytoplasm, is resistant to pepsin. He also found that the Nissl substance contains phosphorus. He concluded that the granules of Nissl substance contain '*nucleoalbumine*' (nucleoprotein). These first observations on the chemical nature of intraneuronal constituents are landmarks in the history of cytochemistry, since they first record the location of nucleic acid and of protein in cellular cytoplasm. Unfortunately Held's remarkable discovery of cytoplasmic nucleic acid was neglected for almost half a century.

Physiological significance of Nissl substance. The relation of Nissl substance to nervous activity was first recognised by Gustav Mann in 1898. He was one of the great pioneers of cytochemistry. His most informative treatise on physiological histology (published in 1902) rationalized the principles underlying the coloration of cellular constituents by dyes. He was one of the great teachers of his time. After taking his M.D. at Edinburgh he held the post of senior university demonstrator at Oxford and later became professor of physiology at Tulane University, New Orleans. Mann showed that the basophilic

Nissl spindles (stainable by toluidin blue) disappear from the spinal anterior horn cells of dogs after excessive muscular exercise. Plate 4 is an illustration from Mann's classical article (1895) on Nissl substance; this figure shows a normal anterior horn cell with copious basophilic spindles and beside it a similar cell body from a fatigued animal, in which the cytoplasm is depleted of most of its basophilic granules. Mann also found that the basophilic substance in the cytoplasm of pyramidal cells in the occipital cortex diminishes after exposure of the eye to light. He concluded that during rest 'chromatic materials are stored up in the nerve cell, and these materials are used up by it during function'. Mann later suggested that the basophilia of both Nissl bodies and of nuclei is dependent on the presence of nucleic acid. Mann's views were confirmed and extended by the work of the great Irish neurologist Gordon Holmes, who is now famous for charting the visual topography of the occipital cortex. Holmes showed in 1903 that Nissl substance disappears from the anterior horn cells of frogs during convulsions induced by strychnine, but, when he cooled the frogs in ice so to prevent the convulsions, then the loss of Nissl substance was arrested although full doses of strychnine were injected. He concluded that the loss of Nissl substance is the *direct result of neuronal 'overwork'* and is not caused by any toxic action of strychnine.

Intraneuronal lipids. Pigmented granules inside neurones were described by von Kölliker (1867, p. 249) in his celebrated textbook of histology. These yellowish granules of wear and tear pigment are usually located beside the nucleus; they tend to become more numerous as age advances. Rosin in 1896 suggested that these pigmented intraneuronal granules contain fatty substance, to which the name *lipofuscin* was later applied. The presence of lipids in the myelin covering 'medullated' nerve fibres was of course appreciated (q.v.) much earlier in the nineteenth century; but lipofuscin was the first lipid to be demonstrated within the neurone. In 1899 John Robert Lord suggested that these yellow granules arise by fatty degeneration of the Nissl substance. Lord indeed showed that these yellow intraneuronal granules are blackened by osmium tetroxide and are partly soluble in ethanol and ether. He thus had cogent reasons for regarding these granules as a product of fatty degeneration. The fatty nature of lipofuscin was finally proved in 1906 by Ernst Sehrt who showed that the pigmented granules are coloured by Sudan III.

Transition to the modern era

By this time then the principal cellular and chemical constituents of the brain had been discovered. The cerebral neurones were regarded as the units of thought, and these complex cells, with their ramifying processes of extreme tenuosity and relatively immense length, were recognized as the integers of cerebral activity and metabolism. The necessity for continuous nutrition of grey matter by the blood was established and even some knowledge existed on the location of substances within nervous cells. The nature of chemical events inside the neurone was, however, still entirely obscure.

By this time also, electrophysiology had made remarkable advances from the preliminary observations of Galvani in the eighteenth century up to the study of action potentials by Einthoven and Keith Lucas in the early twentieth century; by then the conduction of nervous messages was known to be accompanied by equally rapid waves of electrical depolarization. Knowledge of the physical nature of impulses has since been dramatically expanded by the work of Adrian, Matthews, Hodgkin and others in the present century.

As the twentieth century progressed the trail of neurobiochemistry widened and subdivided to such an extent that simultaneous advance in several directions became possible. In the short account which follows only a few of these lines of enquiry can be followed, and these only in briefest summary. The different paths or pursuits of knowledge fortunately at times illuminate and clarify one another; but for the sake of simplicity they are best considered separately. They conveniently fall into two principal groups:

(1) The macromolecular fabric of the neurone. This provides the replicating agents or templates necessary for replenishing the lost molecules of neuronal cytoplasm which are destroyed

by continuous katabolism, and it also constitutes the structural and catalytic framework necessary for the reactions of micromolecules.

(2) The neuronal micromolecules. These furnish the fuel which supports the exergonic reactions necessary for neuronal function as well as for macromolecular synthesis and replenishment. Some of the micromolecules are also the initiators of nervous excitation and transmission.

MACROMOLECULAR FABRIC OF THE NEURONE

Ribonucleic acid of the cytoplasm

Held's shrewd suggestion that Nissl substance is a nucleoprotein remained generally neglected in the first part of the twentieth century. Indeed interest in the chemical nature of the Nissl substance waned as the fashion in biochemistry veered away from living cells to the study of hashes and 'breis'. But in 1932 Lárus Einarson introduced his elegant gallocyaninchromalum method for detection of cellular nucleic acids and he reaffirmed the presence of nucleic acid in the Nissl substance. Although Einarson (1932, 1933) published several important papers on the presence of nucleins in Nissl substance, his work also received little attention at the time.

By 1940, however, interest had become reorientated towards intracellular chemistry. In this year Jean Brachet finally dispelled all doubt about the presence of nucleins in the Nissl substance by the use of ribonuclease for demonstrating ribonucleic acid (RNA). He found that the power of the Nissl substance to bind basic dyes is abolished by pre-treatment with ribonuclease. He thus established that the Nissl substance contains RNA.

At about this time Caspersson and his colleagues at the Karolinska Institut in Stockholm began systematically to investigate the nucleic acids and proteins of both cytoplasm and nucleus by their capacity to absorb ultraviolet light. These methods depend on the fact that the pyrimidine groups in the nucleic acids selectively absorb radiation with wavelength in the vicinity of 2,600 Å, while the tyrosine and tryptophane groups in the proteins absorb in the region of 2,800 Å. From these elaborate and extensive investigations Caspersson concluded that the synthesis of all protein is dependent on the presence of polynucleotides. He supposed that the deoxyribonucleic acid (DNA) of the nucleus is responsible for the synthesis of chromosomal or genic protein, while the RNA of the cytoplasm is the template on which all cytoplasmic protein is manufactured.

Caspersson and his co-workers also discovered that actively dividing cells and those known to fabricate specific proteins (such as digestive enzymes, hormones and haemoglobin) are particularly rich in cytoplasmic RNA, which is required for the fabrication of protein. Other cells, however, with one exception, contain little cytoplasmic RNA. The exception is the neurone. It soon became apparent from the work of Hydén (1943), who confirmed Brachet's identification of RNA in the Nissl substance, that the Nissl substance is indeed one of the most active zones of protein synthesis in the body.

Hydén made a series of observations on RNA in the Nissl substance during and after activity. He confirmed Gustav Mann's conclusion (made half a century earlier) that muscular exercise depletes the anterior horn cells of their basophilic granules; he further found that RNA simultaneously disappears from the same zones in neuronal cytoplasm. Similarly exposure to prolonged and intense auditory stimuli results in loss of RNA from the neurones of the cochlear ganglion. By beautifully designed experiments Hydén later established that prolonged vestibular stimuli also cause loss of RNA from the cells of the vestibular ganglion and of Deiter's nucleus. Hydén moreover found that RNA is rapidly restored inside the depleted cells during recovery after activity. In fact he fully confirmed the earlier work of both Gustav Mann and of Lárus Einarson, who showed that Nissl substance is used up in activity and refabricated during rest.

Hydén's researches also demonstrated that protein in the Nissl substance, as well as RNA, is depleted during activity and that resynthesis of this protein during recovery depends on prior restoration of RNA; in fact RNA is the synthetic template for the manufacture of protein. In the living neurone RNA is thus continually being destroyed in activity and reformed during rest, and neuronal protein follows a similar pattern of depletion and accumulation. The abundance of RNA in the neurone (Caspersson, 1947) is 'thus simply explained by the fact that... during intense function its protein forming system must be able to replace used proteins very rapidly'.

Desoxyribonucleic acid of the nucleus

Although Feulgen and Rossenbeck introduced their classical method for demonstrating cellular deoxyribonucleic acid (DNA) in 1924, little biochemical attention was directed to the biological role of DNA for some considerable time. Feulgen emphasized that DNA occurs only in the nucleus. Caspersson's researches first suggested that DNA is responsible for the synthesis of nuclear genic protein in mitosis just as RNA is necessary for the manufacture of cytoplasmic protein. In the adult neurone, which is incapable of mitosis, the importance of DNA probably rests in its essential participation in fabricating RNA for which the neurone has intense demands. Caspersson and Hydén also supposed that DNA in the nucleolus-associated chromatin is concerned in the cellular synthesis of RNA which is specially abundant in the nucleolus. As early as 1933 Lárus Einarson made the brilliant suggestion that the chromatic material of the Nissl substance of neurones is formed round the nucleolus and then diffuses out into the cytoplasm.

Although adult neurones no longer divide, they possess the the full diploid content of DNA derived from the complete genic pattern. Cytochemical observations on the neurones of female cats gave the first indication that the DNA of the two X sex chromosomes of females is distributed in a different manner from the DNA in the corresponding single X and Y sex chromosomes of males; this difference in distribution results in the presence of perinucleolar satellites rich in DNA which Barr, Bertram and Lindsay (1950) discovered in female neurones. Work on neuronal DNA thus heralded a fundamental cytological principle.

Some signposts in the progress of neurology

Protein

Albuminous substances were first detected in the brain by Foucroy in 1793. Vauquelin, as we have seen, showed that grey matter is richer in protein than white matter; this was confirmed by Thudichum by extensive quantitative determinations. Held in 1895 (q.v.) utilized the novel cytochemical technique of treatment with pepsin to locate protein within neurones. More recent work on the cytochemical distribution of protein in cerebral grey matter has emphasized the abundance of protein in the dendrites and their branches. Plate 5 shows a pyramidal cell from cerebral cortex coloured by the tetrazo method of Danielli; amino acids with cyclic groups contained in the tissue proteins are demonstrated by this method. It is clear from plate 5 that the cell body, the apical dendrite and its branches are rich in protein (Dixon, 1954) containing these reactive groups. As we shall see later much of this protein may be enzymic.

Neurones in proximity to each other may vary profoundly in their content of protein. In 1916 E. V. Cowdry described chromophilic and chromophobic neurones which differ in their capacity to be coloured by dyes. Einarson and also Hydén later established that chromophilia depends on the concentration of protein and RNA within the cells; they found that chromophilia varies inversely with previous nervous activity and that protein and RNA disappear from neurones because of intense function. Plate 6 shows this disparity between the content of protein (detected by Salazar's tannoferric method) in chromophobic and chromophilic Purkinje cells of the cerebellar cortex.

Axoplasm, rich in protein, is continuously flowing down nerve fibres at the rate of 0·2 mm. per day. This flow of axoplasm was revealed by the work of J. Z. Young (1944) and Weiss and Hiscoe (1948); these workers constricted axons and found that axoplasm is dammed up above the constriction to produce axonal swelling. The axoplasm brings protein to all parts of the fibre. Weiss and Hiscoe calculated (1948) that the rate of supply of protein is sufficient to account for the known rate of formation of ammonia by nerve, on the assumption that this ammonia is formed by deamination of the protein. Such utilization of protein necessitates continuous synthesis of protein in the cell body and dendrites under the influence of RNA. The fabrication of protein required by the axon and its terminals thus takes place in the neighbourhood of the nucleus. Weiss and Hiscoe suggested that a continuous flow of protein from its perinuclear source is necessary to replace the continuous loss of protein in the axon. It seems very likely that the energy required for maintaining axonal activity is ultimately provided at the expense of this axonal protein. This view of flow of axonal protein is a remarkable resuscitation of Willis's notion that an essential nutriment for the nerves is supplied by the brain. Moreover the continuous flow of axoplasm manufactured in the cell body and dendrites explains why the cell body is the essential nutritive centre of the fibres, as postulated by Waller, and also why in Wallerian degeneration the fibre dies when it is severed from its source of nutrient protein.

The lipids of grey matter

Grey matter contains less lipid than white matter; but even grey matter ranks high as a lipid-containing organ. Lipids of grey matter include lecithins and kephalins as well as gangliosides (isolated by Klenk in 1941) and strandin (discovered by Folch in 1951). The latter two substances are sphingolipids which contain sphingosine, the cerebral base originally isolated by Thudichum; it is remarkable how frequently his name is cited first in articles on cerebral chemistry today.

These lipids are dispersed in the membranes of the cell body and dendrites, and, unlike neutral fat, do not form droplets stainable by the Sudan dyes; they may be termed *micellar* lipids. But grey matter also contains an aggregated *globular* lipid which is stainable by the Sudan dyes; this is *lipofuscin* or 'wear and tear pigment'. This substance (q.v.) occurs as prominent yellowish granules in neuronal cell bodies.

Although neuronal lipofuscin was discovered in the nineteenth century, its significance is still obscure. Marinesco (1909, vol. 1, p. 287), in his great book *La cellule nerveuse*, noted that this pigmented lipid becomes more abundant with increasing age.

Some signposts in the progress of neurology

Neurones have no replacements and their metabolic activity is lifelong. It is therefore not surprising that some resistant product (possibly derived from the membranous components of neurones) should accumulate in these confined units of intense metabolism over the lapse of years. These granules of globular lipid may then aggregate as the materials responsible for their dispersion are destroyed; lipofuscin may thus persist as an inert residue derived from the perennial degradation of large quantities of neuronal cytoplasm vastly in excess of the final quantity of the remnant. One property of these intraneuronal granules, which strikingly reveals their intracellular location, is the intense rose-red colour which they develop on treatment with the periodic acid-Schiff technique (Dixon and Herbertson, 1950). Some of these deposits of lipofuscin are acid-fast when stained by the Ziehl–Neelsen method, and also fluoresce strongly in ultraviolet light. In this they resemble the copious masses of intraneuronal lipid which Lárus Einarson (1953) discovered in the neurones of animals deprived of vitamin E. Since then Einarson (1962) suggested that lipofuscin in aging neurones may accumulate owing to inability to utilize vitamin E and consequent autoxidation of lipids. Lipofuscin formed in this way may later be engulfed by lysosomes inside the neurones (Adams, 1965, p. 315).

Lipids of white matter

Vauquelin in 1812 (q.v.) recognized that white matter is richer in lipids than grey matter; this excess of lipids in white matter reflects its higher content of myelinated fibres, which indeed, as Schwann pointed out, are responsible for its white appearance.

Myelin is now known to contain several characteristic lipids including the sphingomyelins and cerebrosides, originally discovered by Thudichum, and also cholesterol in the free (non-esterified) state; in fact these lipids were termed the 'myelin lipids' by R. J. Rossiter and his colleagues who have done so much towards defining the chemical nature of myelin (Johnson, McNabb and Rossiter, 1948). The insulation, required for the highly efficient 'saltatory' type of conduction of nervous impulses which occurs in myelinated fibres, almost certainly depends on these special lipid components of myelin. Thudichum's view, that the physical properties of these substances are of supreme importance to nervous activity, was indeed prophetic. The physical properties of myelin have recently been re-emphasized by C. W. M. Adams, who showed (1958) that normal myelin is a hydrophilic or micellar lipid, but that the droplets of lipid formed in Wallerian degeneration are hydrophobic or globular lipids. Thudichum also emphasized, with his characteristic prescience, that changes in the colloidal state of cerebral lipids may be of great importance in disease.

The lipids of myelin acquired a new significance in 1954 when Betty Ben Geren discovered that the myelin sheaths of peripheral nerves are formed by invagination of axons into Schwann cells with subsequent winding of the Schwann cells as spirally flattened laminae around the axons. Myelin is thus probably composed of compressed cell membranes in almost pure form. Studies on the chemical and optical properties of myelin therefore provided remarkable confirmation of the view that the cell membrane is composed of orientated micellar laminae of lipid with adherent films of protein.

MICROMOLECULES AND THEIR METABOLISM

Nutritive micromolecules

Utilization of oxygen. The dependence of cerebral function on the continuous supply of oxygen to the brain was early recognised. By the mid-nineteenth century William Stokes (q.v.) suggested that cerebral ischaemia causes confusion and coma, while Charles-Edouard Brown-Séquard (q.v.) established the vulnerability of cerebral tissue to anoxia as well as the revivification of anoxic brain by oxygenated blood. M. Litten in 1880 showed that compression of the abdominal aorta in rabbits for a period of 1 hr. is followed by permanent paralysis of the lower limbs and incontinence of urine; Litten suggested that these disabilities result from ischaemic injury to the spinal cord, and he noted that even the kidneys, which are themselves more susceptible to ischaemic damage than most tissues, are not so rapidly injured by ischaemia as the spinal cord. In 1884 Ehrlich and Brieger followed up Litten's observations and showed that this compression of the abdominal aorta in fact kills the anterior horn cells of the spinal cord. Ehrlich and Brieger also found that the grey matter of ischaemic spinal cord shows microscopic evidence of injury sooner than the white matter.

These observations implied that the neurone is highly vulnerable to anoxia and that the cell body and dendrites are more susceptible than the axon. In a masterly paper published in 1885 Paul Ehrlich emphasized the intense reducing activity of cerebral grey matter (*v.* Ehrlich, 1956): he was able to show that living grey matter rapidly reduces alizarin blue to a colourless compound and that dead grey matter in the spinal cord, which has been killed by ischaemia, fails to effect this reduction (Ehrlich and Brieger, 1884). Ehrlich realized that grey matter requires oxygen and he further established that brain absorbs oxygen and uses it vitally. To demonstrate this utilization of oxygen he injected a mixture of dimethyl-paraphenylenedi-amine and α-naphthol intravenously into rabbits and found that these substances are *oxidized* and condensed by brain tissue to form indophenol blue which is deposited as dark blue granules.

Ehrlich gave the following equation to symbolize his concept of 'vitale indophenolsynthese':

$$C_6H_4 \begin{cases} NH_2 \\ N(CH_3)_2 \end{cases} + C_{10}H_7[\alpha]OH + O_2$$

$$= C_6H_4 \begin{cases} [1]N(CH_3)_2 \\ [4]N{=}C_{10}H_5[\alpha]OH \end{cases} + 2H_2O$$

He showed that the brain and the heart possess to a paramount extent this power of intravital oxidation.

The capacity of central nervous tissue directly to consume gaseous oxygen was first demonstrated by Hans Winterstein in 1906. He showed (1906, 1907) that isolated spinal cords of frogs absorb oxygen and that this consumption of oxygen is increased by electrical stimulation. Winterstein's work is notable for his use of Thunberg's microrespirometer for the study of respiration in nervous tissue.

A most thorough study of neuronal lesions produced by anoxia was made by the great Roumanian neuropathologist Gheorghe Marinesco from 1896 onwards. In Marinesco's *La cellule nerveuse*, published in 1909, he stressed the special susceptibility of grey matter to damage by anoxia, which he related to the intense utilization of oxygen necessary for maintaining both the structure and the function of nervous tissue. Marinesco demonstrated (vol. II, p. 429–64) profound loss of Nissl substance from the periphery of the cell bodies (peripheral chromatolysis) in spinal anterior horn cells of rabbits within 3 hours of ligature of the abdominal aorta. The power of neurones to absorb oxygen by cellular respiration and their dependence on continuous supply of oxygen were thus fully recognized by the early twentieth century. It was later established that slices of cerebral cortex absorb oxygen when suspended in Ringer's solution containing oxidizable substrates (Warburg, Posener and Negelein, 1924).

Utilization of sugar and glycolysis. In 1917 Hans Winterstein in collaboration with Else Hirschberg discovered the power of nervous tissue to utilize glucose. Hirschberg and Winterstein demonstrated that added glucose is lost from the solution in which the spinal cords of frogs are incubated; these workers also found that the consumption of glucose is increased by electrical stimulation and diminished by the addition of ethanol or urethane. This pioneer work on the utilization of glucose is often neglected in recent discussions on this subject.

In 1922 F. C. Mann and T. B. Magath showed that dogs lose consciousness after removal of the liver owing to rapid depletion of glucose from the blood; moreover consciousness was found to be restored by administration of glucose. In 1923 J. J. R. MacLeod discovered that injection of insulin may cause similar hypoglycaemic coma. It was thus realized that glucose is an essential nutriment for the brain. The continuous utilization of glucose by human brain during life was demonstrated in a particularly elegant manner by Harold Himwich and his colleagues in 1939 (Himwich, 1951, p. 13). They used the cerebral 'arteriovenous oxygen difference' to assess the utilization of oxygen. This difference is lowered by hypoglycaemia and is restored by feeding glucose. From this they most reasonably concluded that glucose is the principal nutriment of the living brain.

In 1924 Otto Warburg and his colleagues discovered that slices of cerebral cortex of rats, in the absence of oxygen, rapidly convert glucose to lactic acid, and that this glycolysis is markedly diminished by oxygen. The effect of oxygen in diminishing the katabolism of glucose by cerebral cortex was later demonstrated by direct measurements of the destruction of glucose by slices of cerebral cortex. Grey matter thus displays the Pasteur Effect; this is a widely possessed property of living tissues, by which oxygen diminishes the consumption of carbohydrate and so prevents needless loss of nutriments. The accelerated glycolysis of cerebral cortex in the absence of oxygen has been likened to a vain attempt to gain sufficient free energy from the uneconomic process of anaerobic glycolysis.

Glucose is thus the main fuel of the brain. During transient anoxia glycolysis to lactic acid may temporarily provide sufficient free energy for maintaining neuronal life, though the energy made available by this anaerobic fission of glucose is probably insufficient for nervous activity to continue. For this reason it was suggested that the normal process of consciousness depends on the rate of liberation of free energy in the cerebral neurones attaining *a certain level*; below this level coma supervenes (Dixon, 1937). Cerebral ischaemia in man and animals rapidly causes loss of consciousness in as short a time as seven seconds, but irreparable damage to cerebral neurones usually does not occur until ischaemia has persisted for a few minutes (Rossen *et al.*, 1943; Weinberger, 1940*a*, *b*). This short period of immunity to permanent damage may depend on the continued supply of energy at a lower rate by the anaerobic fission of glucose still present in the tissues; thus anaerobic glycolysis may *very temporarily* provide a rate of energy supply high enough to prevent irreversible injury but not to maintain nervous activity. Cerebral cortex is damaged irreversibly by relatively short periods of ischaemia, while lower centres, such as the medulla, can survive for longer periods of temporary interruption in their supply of blood (Weinberger *et al.*, 1940*a*, *b*). The extreme dependence of cerebral cortex on continuous supply of both oxygen and glucose may be related to the high rate of cortical glycolysis and may thus reflect the high demands of energy required by this uniquely vulnerable tissue.

The actual steps in the oxidation of glucose by cerebral tissue to CO_2 and water were revealed by the illuminating work of Rudolph Peters and his colleagues (from 1929 onwards). Peters discovered that pyruvic acid is an intermediary in this process and that deficiency in vitamin B_1 interrupts the further utilization of pyruvic acid which then accumulates. Peters also showed that poisoning by fluoracetate breaks the chain of reactions at a later stage so that citric acid, which is formed by the oxidative metabolism of pyruvic acid, is then debarred from further metabolism; in animals poisoned by fluoracetate citrate therefore accumulates and was detected in large quantities in the brain (Buffa and Peters, 1949). Peters thus established that the citric acid cycle of oxidation is an integral part of the aerobic katabolism of carbohydrate by cerebral tissue.

Location of respiration and glycolysis. The intraneuronal sites of micromolecular metabolism are determined by the intracellular disposition of enzymes which form part of the macromolecular fabric of the cell. Paul Ehrlich's demonstration in 1885 (q.v.) of the synthesis of indophenol blue in cerebral tissue gave the first insight on the location of vital catalysts in brain. H. M. Vernon, in 1911, showed that this oxidation is due to the enzyme 'indophenol oxidase' which is specially abundant in grey matter and is also present in other actively respiring tissues. D. Keilin later established that the enzyme responsible for the bluing of Ehrlich's indophenol reagent is in fact cytochrome oxidase. This enzyme oxidizes reduced cytochrome to oxidized cytochrome and the latter substance then oxidizes the indophenol reagent to the quinonoid indophenol blue. The active reduction of methylene blue by cerebral grey matter was discovered by C. A. Herter in 1905, though Paul Ehrlich clearly appreciated the power of brain to reduce several coloured dyestuffs (including methylene blue) to their leuco-bases.

In 1912, Siegfried Gräff found that the granules of indophenol blue formed by the action of indophenol oxidase are deposited in the *cytoplasm* of the ganglion cells and not in the nuclei. He also showed that death of neurones from infarction causes loss of the oxidizing granules. Finally in 1919, Gheorghe Marinesco gave a full description of the location of indophenol oxidase in the neurone. He showed that all grey matter is selectively delineated by the indophenol reaction. Sharply defined grey matter, such as that in the inferior olive, stands out against a colourless background of unreactive white matter when treated *in vitro* with Ehrlich's indophenol reagent; in fact, as Marinesco said, the grey matter is signalized with 'une clarité étonnante'. He also showed that this striking differentiation is due to the presence of indophenol oxidase in the cell bodies and dendrites and its absence from the axons. Marinesco's illuminating paper indicated for the first time the dominance of the dendrites in cerebral metabolism. Marinesco was also the first to suggest that indophenol oxidase is located in the mitochondria: he was thus one of the prime founders of neuronal cytochemistry (Petrescu, 1964).

Following pioneer work of Hans Winterstein and of Otto Warburg on the metabolism of isolated tissues much research was devoted to the metabolic behaviour of sliced and chopped brain suspended in Ringer's solution containing glucose. E. G. Holmes (1930) showed that grey matter consumes oxygen and forms lactic acid much more rapidly than white matter. The question then arose as to which part of grey matter is responsible for this high metabolic activity: is this located in the cell bodies or in the dendrites?

Owing to the extreme length of many neurones it is possible to study the metabolism of one part of the cell alone. This principle was first applied by E. G. Holmes in 1932. He showed that the Gasserian ganglion (which contains cell bodies but no synapses nor dendrites) behaves like white matter and has low rates of respiration and glycolysis; he concluded therefore that the high metabolic activity of cerebral grey matter is due to reactions in the dendrites or synapses.

But it could be contended that sensory ganglia (like the Gasserian) are a form of grey matter so dissimilar from the higher centres that deductions made from observations on ganglionic metabolism could not validly be applied to the behaviour of cerebral cortex. However in cerebral cortex it is possible to measure metabolic activity of tissue containing the apical dendrites without the participation of the majority of their cell bodies. This is achieved by cutting tangential slices from the cerebral surface. These slices are largely composed of the plumate dendritic processes of pyramidal cells (and terminal fibres) and contain relatively few cell bodies. The rate of glycolysis of these superficial slices was found to be no less but actually slightly greater than that of deeper layers which contain neuronal cell bodies and nuclei in abundance (Dixon, 1953). It is likely therefore that glycolysis, like respiration, is mainly located in the dendrites and synapses which form the dominant metabolic components of the brain. The abundant protein within the dendrites and their branches is probably rich in enzymes, which are concerned with the numerous reactions of dendritic metabolism. Studies on the distribution of individual enzymes in the different layers of hippocampal cortex (Lowry *et al.*, 1954) and of visual cortex and motor cortex (Robins *et al.*, 1956) also indicate the large contribution of the dendrites to cerebral metabolism as a whole.

Excitatory and inhibitory substances

Synaptic transmitters. The action of these substances were discovered in the present century by the work of T. R. Elliott, O. Loewi, H. Dale, J. C. Eccles and others. These specific micromolecules are liberated at synapses and there excite or inhibit other neurones. These same substances are also liberated at axonal terminals in various organs where they are similarly responsible for transmitting either activity or inhibition from neurones to other types of cells. These transmitters include acetylcholine, adrenaline and noradrenaline. Recent work suggests that 5-hydroxytryptamine, which is also present in cerebral tissue, may belong to this class of hormones with specific action on nervous tissue.

Substances formed inside the body may thus exert highly specific effects on the neurone in its normal function. It is also possible that such substances may cause neuronal *malfunction* and so become responsible for mental disorder. In 1952 Osmond and Smythies showed that the drug mescaline can produce intoxication closely resembling schizophrenia. Osmond and his colleagues (Hoffer *et al.*, 1954) also discovered that adreno-chrome, a quinone formed from adrenaline by oxidation, produces hallucinations very similar to those caused by mesca-line. Abnormal metabolism within the body might easily oxidize adrenaline to adrenochrome. Osmond very reasonably suggested that adrenochrome may thus be a naturally occurring 'hallucinogen', which can be formed in the body by deviation of normal metabolism. The possibility that mental diseases are determined by metabolic defects along with the accumula-tion of harmful products has therefore become a challenging signpost for research into the hitherto inscrutable aberrations of the mind; this view was first envisaged by Thudichum (see fig. 4).

Impulse-propagating substances. In 1949 the brilliant work of A. L. Hodgkin and his colleagues (1951) revealed the prime importance of ionic movements in causing and propagating the electrical changes of the nerve impulse. The entry of sodium and the exit of potassium are basic determinants of the waves of depolarization and recovery which constitute nervous conduc-tion. Later sodium is expelled and potassium re-enters so that the previous ionic dispositions are restored; this final restora-tion requires energy supplied by metabolism which is also necessary for preserving the polarized state of the resting cell membrane. It seems likely that in myelinated fibres these ionic changes only occur at the nodes of Ranvier and that depolariza-tion at one node results in saltatory conduction of the impulse to the next node. The insulating properties of the lipids of myelin are essential for this highly efficient form of conduction.

The mechanism by which expenditure of energy is used to

maintain ionic gradients and membrane potentials is still obscure. However there is considerable evidence that compounds containing high-energy phosphate bonds are involved as immediate stores of potential energy.* The work of P. J. Heald (1960, pp. 117–24) has been particularly fruitful in directing attention to the possible participation of phosphoprotein in restoring the polarized state and to the linkage of this process of maintenance and restoration with the energy supplied by respiration and glycolysis. The dependence of high-energy phosphate bonds on the continuous supply of nutrients has also clearly been emphasized by M. Schneider (1957), who found that phosphocreatine and adenosinetriphosphate are lost from brain after quite short periods of ischaemia.

METABOLISM IN RELATION TO FUNCTION

Chemical requirements of the active neurone

A biochemical vista of the active neurone, although still nebulous in detail, has thus begun to clarify in at least two well-defined features. To persist in the living state the neurone requires (i) a permanent and self replicating macromolecular pattern associated with the capacity for exceptionally rapid synthesis of protein, and (ii) a constant supply of nutrient micromolecules. We shall now briefly consider why these two requirements are of such unique importance to the continuity of neuronal life and function.

Permanent and self-replicating macromolecular pattern. Since neurones are no longer capable of division, replication and synthesis within each of them must be *lifelong*. This intracellular replication provides the RNA required to fabricate the abundant protein utilized by the neurone. Protein is, in the first place, needed to provide for the continuous centrifugal flow of axoplasm. Weiss and Hiscoe, as we have seen, suggested that this axonal protein is the source of ammonia formed by nerve; thus protein may provide the energy for maintaining peripheral

---

* Recent information on this subject is given by S. Rose (1965).

nerve. Another kind of protein elaborated by neurones is the neurosecretory substance, which originates in the hypothalamus. The work of E. and B. Scharrer and W. von Bargmann indicated that the active principles of the posterior lobe of the pituitary gland originate from this product of hypothalamic neurosecretion. C. W. M. Adams and J. C. Sloper (1956) showed that this substance occurs in the cell bodies of human hypothalamic neurones and that it is a protein rich in cystine. In addition to these special nutrient proteins, the neurone, like other cells, contains its complement of enzymic and structural proteins, which are also fabricated by the abundant RNA present in the cell bodies and dendrites.

Constant supply of nutrient micromolecules. Continuous supply of glucose and oxygen provides for the intense glycolysis and respiration of grey matter. The free energy liberated in these exergonic reactions is responsible for the synthesis of adenosine triphosphate (ATP) and other compounds containing high-energy phosphate bonds, which in their turn sustain the synthesis of protein and the fabrication of RNA.

Glycolysis and respiration may also provide energy for restoring neuronal ionic disposition after activity: in this they probably act through the mediation of ATP and other compounds containing high-energy phosphate bonds (Caldwell *et al.*, 1960). We have seen that sodium ions enter nerve and that potassium ions are released during the passage of impulses. If similar changes occur in brain during activity, the concentration of potassium ions in the fluid outside the neurones should rise and at the same time the concentration of sodium ions should decrease. Indeed precisely these changes in ionic environment are known to exert profound effects on cerebral metabolism by stimulating both aerobic glycolysis and respiration (Ashford & Dixon, 1935; Dixon, 1949). This remarkable action of potassium ions in accelerating aerobic glycolysis and respiration led to the suggestion that waves of increased metabolism, induced by liberated potassium ions, may be a normal concomitant of central reflex activity (Dixon, 1940). Electrical stimulation of grey matter also increases glycolysis and respiration (Winterstein, 1906; Hirschberg and Winterstein, 1917; Klein and Olsen, 1947; McIlwain *et al.*, 1951). Moreover it is now believed that this increase may be caused by potassium ions which are liberated by electrical stimulation (Cummins and McIlwain, 1961); in fact the potassium liberated by the application of electric pulses may stimulate metabolism and thus provide energy necessary for the re-entry of the extruded potassium. It is thus reasonable to return to the concept that glycolysis and respiration, which have been augmented by potassium ions liberated during *normal activity* (Dixon, 1940), constitute an important portion of active cerebral metabolism; these reactions stimulated by potassium ions may indeed provide a natural 'feed back' mechanism by which extra energy is supplied after activity for expelling sodium ions from the neurones and for restoring the internal concentration of potassium. Thus the substantial effects of potassium ions on the metabolism of cerebral tissue, which have been revealed by experiments on tissue slices, may reflect metabolic adjustments of considerable magnitude which are constantly occurring within the living active neurone.

Under anaerobic conditions, on the other hand, potassium ions exert a rapid and profound inhibitory effect on the glycolysis of slices of cerebral cortex. Local excess of potassium may exist around ischaemic neurones owing to previous nervous activity or because of tissue damage or haemorrhage. Under these conditions the ischaemic neurones would then be deprived of their principal residual source of energy provided by anaerobic glycolysis, which is itself so substantially inhibited by potassium. The unique susceptibility of cerebral neurones to damage by ischaemia may thus in part depend on the inhibitory action of potassium ions on cortical anaerobic glycolysis as well as on the cessation of oxidative phosphorylation in a tissue which has so high a demand for continuous supply of energy.

Some signposts in the progress of neurology

Cerebral fabric and thought

Electrical activity in the brain of animals was first discovered*
by Richard Caton in 1875; he described spontaneous activity
unrelated to external stimuli as well as localized potential
changes correlated with sensory and motor phenomena. Hans
Berger in 1929 first drew attention to rhythmic changes of
potential in human cerebral cortex. The elegant work of
E. D. Adrian then established that electrical activity in human
cortex is related to the process of mental attention and thought.

More recently Lashley (1941) showed that visual concepts,
such as the mental recognition of triangles, are formulated in
the visual cortex; he suggested that thought consists of resonance
or interference patterns in innumerable local circuits made by
cortical cells and their conducting elements. K. W. Craik (1943)
further developed this view, he considered that symbolic
models of physical reality are the basic contents of thought,
and that these small-scale models are formed by patterns of
electrical activity in the brain. This kind of scheme, however,
is not very different from Locke's theory that our ideas only
represent external substance or indeed from Kant's concept of
phenomena. The latter Kant supposed merely represent within us
the *noumena* of the real world; these noumena are the *things in
themselves*, which we never observe directly but only through a
long chain of perceptual processes.

To the biochemist a molecular basis for these patterns of
electrical activity would appear almost axiomatic for the
permanence of thought and memory. Craik in his book on the
Nature of explanation realized this difficulty 'that some more
permanent pattern depending possibly on synaptic resistance. . .
must be postulated to account for memory'. This difficulty in
accounting for the persistence of memory is inherent in any
hypothesis that thought depends on mere patterns of electrical
activity.

Adrian (1947), in his *Physical background of perception*, pointed

* A valuable account of early work on electrical activity of brain was published
by Mary A. B. Brazier (1961) *A history of the electrical activity of the brain: the
first half century.* (London), pp. 4–10.

out this difficulty very forcibly. He wrote 'if memory traces are patterns of activity, the patterns must tend to recur after a period of inactivity'. In fact he suggested 'some kind of material change is needed to account for persistence' after interruptions as in deep anaesthesia. However Adrian still accepted 'the hypothesis that the physical basis of memory is in the nature of a resonance pattern...established in local circuits throughout the whole cortex'. Surely this cannot be all, especially when the records of memory are so permanent and so perennial.

Transient and persistent memory. After cerebral concussion, which causes coma with a prolonged nervous refractory period and interruption of electrical activity, there is commonly *retrograde amnesia* for incidents immediately preceding the accident. The most recent events are in fact less indelibly impressed on memory and may be erased; but the more remote events are usually quite unaltered and permanent so as to be recaptured completely once consciousness returns. A similar permanence of memory for long past events is evident in old age when recent happenings begin to fade.

There thus appear to be two degrees of memory, one transient, the other perennial; once the permanent level is reached by any idea it is apprently permanently impressed on the records of the mind.

Macromolecular fabric and memory. If a mere pattern of electrical activity were the entire basis of thought, then following the coma of concussion, all previous thoughts should die, never to return. We must therefore postulate some change impressed on the macromolecular mosaic of the neurone, which is after all 'the most diversified chemical laboratory in the body'— to use Thudichum's metaphor. The macromolecules of proteins and lipids constitute an assembly of variable pattern complex enough to account for an almost infinite catalogue of thoughts. The pattern of local depolarization at neuronal surface membranes, upon which electrical activity depends, may thus be based on variation in the constitution and arrangement of lipid and protein molecules. Incoming impulses may produce transient changes in the orientated lipids of the neuronal surface; these changed patterns may then cause specific adsorption and alignment of complementary proteins or peptides from within the cell. These complementary peptide patterns may be able to remould the same new lipid patterns on the surface and thus regenerate the same patterns of electrical activity as were originated by the new impression. Thus a pattern of electrical activity would become permenent if a complementary inherited peptide was already present in the cell. Such peptides may only be present originally in minute amount; their synthesis may initially be suppressed by their effector action on regulator genes, which specifically block the activity of their complementary DNA in the genomic code. Removal of a peptide pattern by adsorption on a complementary lipid pattern at the cell surface would then arrest its suppressor action; thereafter the complementary part of the genomic code would be unmasked with attendant formation of the corresponding messenger RNA and subsequent fabrication of this same peptide. Thus the mechanism for synthesis of a new peptide pattern and the remoulding of new lipid and electrical patterns would become catalogued in the archives of permanent memory and thought.*

This view accepts that electrical activity is essential for the process of thought, but supposes that thought and memory are based on more than a mere pattern of impulses. It seems more likely that there are underlying patterns impressed on the macromolecular fabric of the neurones. Such a macromolecular mosaic could regenerate electrical patterns; even though these patterns have been temporarily eliminated they can later recur under the dominant influence of an unmasked portion of the

---

* J. Katz and W. C. Halstead (1950, *Comp. Psychol. Monogr.* 20, 1–38) suggested that memory may depend on structure of cerebral proteins. The view that a mosaic of protein and lipid (depending on complementary RNA for its reformation at neuronal surfaces) is concerned with memory and concussion was advanced in this lecture given at Cambridge in 1958 (Dixon, K. C. *Lancet*, 1962, Dec. 29, 1359–60). H. Hyden (1958, Symposium no. 3 of 4th International Congress of Biochemistry, 83) expressed somewhat similar views. Moreover P. Glees (1962, *die Umschau in Wissenschaft und Technik.* 14, 435–8) produced cogent evidence for a molecular basis of memory. The possible intervention of DNA in this kind of mechanism was suggested more recently (Dixon, K. C. *Lancet*, 1967, Jan. 7, 27–8).

genetic code. The transience of *recent* memory traces (which may be eradicated by concussion) implies that the synthetic mechanism needed for the continued reformation of each new pattern of lipid and protein is not immediately triggered into action. But once unmasked each portion of the genetic DNA may continue to fabricate complementary RNA so that the new configurations of protein and lipid become part of the permanent patterns of the cell. This would account for the permanence of memory and also for that most conspicuous feature of neuronal metabolism, namely the formidable capacity for intense synthesis of protein and its all-important replenishing fabricator RNA. Only when the basic templates are decomposed, so as to prohibit further fabrication of their specific proteins, does memory finally and irreparably perish.

Conclusion

Some biochemical and histochemical characteristics of neurones and their house the brain have thus become apparent during several centuries of human thought and effort. These studies have illuminated two principal kinds of cerebral components.

(1) *The small molecules supplied to the brain by the blood.* These provide the energy essential for the life and activity of the neurone; they also furnish the units for replenishment of the macromolecular fabric and for the synthesis of transmitter substances.

(2) *The macromolecules.* These, in almost infinite variety, form the dynamic edifice on which the patterns of neuronal electrical activity and human thought and memory depend. These macromolecules include (*a*) the *lipids* which form the orientated patterns necessary for varied electrical activity and insulation, (*b*) the *proteins*—enzymic, structural, and nutritive, and (*c*) a third group of substances, which are possibly the most discriminating of all bodily fabricators—the *nucleic acids* of the neurone; these last are responsible for the perennial 'renouvellement' of neuronal substance which was so brilliantly revealed by Augustus Waller.

Such a view of cerebral composition and function in a sense resembles that of Thomas Willis; he held that the spirits of the

Some signposts in the progress of neurology

mind are procreated from the blood 'in the brain, which is their principal shop or workhouse' and from which they are 'newly brought forth' having there been 'kept as it were in distinct cloisters of cells, to be drawn forth for the manifold exercises' of mental function (Willis, VI, p. 95).

As investigation uncovers and unravels more neuronal complexities, there is displayed an ever receding barrier, backed by an endless unknown. It is, however, salutary to realize that in spite of continuous and at times spectacular progress, we know so little about the intimate mechanisms of the mind and its dwelling place the brain. It may be that the nature of the human mind is such, that nothing can ever be seen by it or of it more clearly than through the glass darkly of St Paul, or by means of the distorting medium of the forms of human intuition and the categories so vividly imagined by Immanuel Kant.

The search, however, gives zest and possibly the process of search may be our nearest approach to reality. The criticism of the over-satisfied followers of Kepler and Newton by Alexander Pope is at present particularly applicable to any deductions from mental models or electronic brains which are too confident in our capacity to explain ourselves. We would do well to remember Pope's summarizing words:

> Could he whose rules the rapid Comet bind,
> Describe or fix one movement of his mind.

These words are as true today as they were in 1734.

REFERENCES

Adams, C. W. M. (1958). Histochemical mechanisms of the Marchi reaction for degenerating myelin. *J. Neurochem.* **2,** 178–86.

Adams, C. W. M. (1965). *Neurohistochemistry.* Amsterdam.

Adams, C. W. M. & Sloper, J. C. (1956). The hypothalamic elaboration of posterior pituitary principles in man, the rat and dog. Histochemical evidence derived from a performic acid-alcian blue reaction for cystine. *J. endocrinol.* **13,** 221–8.

Ashford, C. A. & Dixon, K. C. (1935). The effect of potassium on the glycolysis of brain tissue with reference to the Pasteur effect. *Biochem. J.* **29,** 157–68.

Kendal Dixon

Barr, M. L., Bertram, L. F. & Lindsay, H. A. (1950). The morphology of the nerve cell nucleus, according to sex. *Anat. Rec.* **107**, 283–92.

Bell, Charles (1811). *Idea of a new anatomy of the brain*. London.

Bell, Charles Sir (1870) *Letters of Sir Charles Bell selected from his correspondence with his brother George Joseph Bell*. London.

Brachet, J. (1940). La détection histochimique des acides pentosenucléiques. *Compt. rend. Soc. Biol.* (Paris) **133**, 88–90.

Brazier, M. A. B. (1961). *A history of the electrical activity of the brain: the first half century*. London.

Brown-Séquard, C. E. (1858). Recherches expérimentales sur les propriétés physiologiques et les usages du sang rouge et du sang noir et de leurs principaux éléments gazeux, l'oxygène et l'acide carbonique. *Journal de la physiologie de l'homme et des animaux.* **1**, 95–122, 353–67, 729–35.

Buffa, P. & Peters, R. A. (1949). The *in vivo* formation of citrate induced by fluoracetate and its significance. *J. Physiol.* **110**, 488–500.

y Cajal, S. Ramon (1894). *Les nouvelles idées sur la structure du système nerveux chez l'homme et chez les vertébrés*. Édition française revue et augmentée par l'auteur, traduite de l'espagnol par le Dr L. Azoulay, Paris.

y Cajal, S. Ramon (1928). *Degeneration and regeneration of the nervous system*. Vol. 1. London.

Caldwell, P. C., Hodgkin, A. L., Keynes, R. D. & Shaw, T. I. (1960). The effects of injecting energy-rich phosphate compounds on the active transport of ions in the giant axons of *Loligo*. *J. Physiol.* **152**, 561–90.

Caspersson, T. (1947). 'The relations between nucleic acid and protein synthesis' in *Symposia of the Society for Experimental Biology, No. 1., Nucleic acid*. Eds. G. H. Bourne and J. F. Danielli. Cambridge University Press.

Cowdry, E. V. (1916). The structure of chromophile cells of the nervous system. *Contrib. embryol. Carnegie Inst.* **4**, 27–43.

Cummins, J. T. & McIlwain, H. (1961). Electrical pulses and the potassium and other ions of isolated cerebral tissues. *Biochem. J.* **79**, 330–41.

Descartes, R. (1650). *Les passions de l'âme*.

Descartes, R. (1664). *Traité de l'homme*. Paris.

Dixon, K. C. (1937). The carbohydrate metabolism of the cerebral cortex. *Irish J. mec. Sci.* 6th series, no. 13, 220–2.

Dixon, K. C. (1940). A study of cortical metabolism in relation to cerebral disease. *Brain.* **63**, 191–9.

Some signposts in the progress of neurology

Dixon, K. C. (1949). Action of potassium ions on brain metabolism. *J. Physiol.* **110**, 87–97.

Dixon, K. C. (1953). Glycolysis and cytochemistry of cerebral cortex. *J. Physiol.* **120**, 267–77.

Dixon, K. C. (1954). Cytochemistry of cerebral grey matter. *Quart. J. exper. Physiol.* **39**, 129–50.

Dixon, K. C. & Herbertson, B. M. (1950). Clusters of granules in human neurones. *J. Path. Bact.* **62**, 335–9.

Drabkin, D. L. (1958). *Thudichum: chemist of the brain*, Philadelphia.

Ehrlich, P. (1956). 'The requirement of the organism for oxygen' in *Collected papers of Paul Ehrlich*. Ed. F. Himmelweit. London. Vol. I, pp. 433–96.

Ehrlich, P. & Brieger, L. (1884). Über die Ausschaltung des Lendenmarkgrau. *Z. klin. Med.* Supplement to Vol. 7, 155–64.

Einarson, L. (1932). A method for progressive selective staining of Nissl and nuclear substance. *Amer. J. Pathol.* **8**, 295–305.

Einarson, L. (1933). Notes on the morphology of the chromophil material of nerve cells and its relations to nuclear substances. *Amer. J. Anat.* **53**, 141–75.

Einarson, L. (1953). Deposits of fluorescent acid-fast products in the nervous system and skeletal muscles of adult rats with chronic vitamin E deficiency. *J. Neurol. Neurosurg. Psychiat.* **16**, 98–109.

Einarson, L. (1962). 'Aging round the world' in *Biological aspects of aging, International Association of Gerontology*. Vol. III, N. W. Shock (ed.) New York. p. 141.

Fourcroy, A. F. (1793). Examen chimique du cerveau de plusieurs animaux. *Ann. Chim. Phys.* **16**, 282–322.

Geren, B. B. (1954). The formation from the Schwann cell surface of myelin in the peripheral nerves of chick embryos. *Exper. Cell Res.* **7**, 558–62.

Gräff, S. (1912). Die Naphtholblau-Oxydasereaktion der Gewebszellen nach Untersuchungen am unfixierten Präparat. *Frankfurter Zeitschrift für Pathologie.* **11**, 358–84.

Guthrie, D. (1945). *A history of medicine.* London.

von Haller, Albrecht. (1757–66). *Elementa physiologicae corporis humani*. Vol. I–IV. Lausanne.

Heald, P. J. (1960). *Phosphorus metabolism of brain.* Oxford.

Held, H. (1895). Beitrage zur Struktur der Nervenzellen und ihrer Forstätze. *Archiv. für Anat. u. Physiol. Anat. Abtheilung*, pp. 396–416.

Herter, C. A. (1905). On the reducing action of the animal organism under the influence of cold. *Amer. J. Physiol.* **12**, 128–38.

Himwich, H. E. (1951). Brain metabolism and cerebral disorders. Baltimore.

Kendal Dixon

Hirschberg, E. & Winterstein, H. (1917). Uber den Zuckerstoffwechsel der nervosen Zentralorgane. *Z. physiol. Chem.* **100**, 185–202.

Hodgkin, A. L. (1951). The ionic basis of electrical activity in nerve and muscle. *Biol. Rev.* **26**, 339–409.

Hoffer, A., Osmond, H. & Smythies, J. (1954). Schizophrenia: a new approach. II. Result of a year's research. *J. mental science.* **100**, 29–45.

Holmes, E. G. (1930). Oxidations in central and peripheral nervous tissue. *Biochem. J.* **24**, 914–25.

Holmes, E. G. (1932). The metabolic activity of the cells of the trigeminal ganglion. *Biochem. J.* **26**, 2005–9.

Holmes, G. (1903). On morphological changes in exhausted ganglion cells. *Z. allg. Physiol.* **2**, 502–15.

Hydén, J. (1943). Protein metabolism in the nerve cell during growth and function. *Acta Physiol. Scand.* **6**, suppl. 17.

Johnson, A. C., McNabb, A. R. & Rossiter, R. J. (1948). The lipids of normal brain. *Biochem. J.* **43**, 573–7.

Klein, J. R. & Olsen, N. S. (1947). Effect of convulsive activity upon the concentration of brain glucose, glycogen, lactate and phosphates. *J. Biol. Chem.* **167**, 747–56.

von Kölliker, A. (1867). *Handbuch der Gewebelehre des Menschens für Aerzte und Studierender*, Leipzig.

Lea, A. Sheridan. (1892). The chemical basis of the animal body: an appendix to *Foster's text book of physiology*. London, p. 137.

Litten, M. (1880). Untersuchungen über den haemorrhagischen Infarct und über Einwirkung arterieller Anaemie auf das lebende Gewebe. *Z. klin. Med.* **1**, 131–227.

Lord, J. R. (1898). A new Nissl method. Normal cell structure and cytological changes terminating in fatty degeneration. *J. mental Sci.* **44**, 693–700.

Lowry, O. H., Roberts, N. R., Leiner, K. Y., Wu, M.-L., Farr, A. C. & Albers, R. W. (1954). The quantitative histochemistry of brain. III. Ammon's horn. *J. biol. Chem.* **207**, 39–49.

McIlwain, H., Anguiano, G. & Cheshire, J. D. (1951). Electrical stimulation of metabolism of glucose by mammalian cerebral cortex. *Biochem. J.* **50**, 12–18.

Mann, G. (1895). Histological changes induced in sympathetic, motor and sensory nerve cells by functional activity. *J. Anat. and Physiol.* **29**, 100–8.

Marinesco, G. (1909). *La cellule nerveuse.* Vols. I and II. Paris.

Marinesco, G. (1919). Recherches sur les oxidases. *Compt. rend. Soc. Biol.* **82**, 98–102.

Some signposts in the progress of neurology

Nissl, F. (1892). Über die Veränderung der Ganglienzellen am Facialskern des Kannichens nach Ausreissung der Nerven. *Allg. z. Psychiatr.* **48**, 197–8.

Noback, C. R. & Montagna, W. (1952). Histochemical studies of the myelin sheath and its fragmentation products during the Wallerian degeneration. *J. comp. Neurol.* **97**, 211–32.

Osmond, H. & Smythies, J. (1952). Schizophrenia: a new approach. *J. mental Sci.* **98**, 309–15.

Petrescu, A. (1964). Gheorghe Marinesco 1863–1938. Promoteur de l'histochimie du système nerveux. *Ann Histochim.* **9**, 145–53.

Robins, E., Smith, D. E., Eydt, K. M. & McCaman, R. E. (1956). The quantitative histochemistry of the cerebral cortex. II. Architectonic distribution of nine enzymes in the motor and visual cortices. *J. Neurochem.* **1**, 68–76.

Rose, S. (1965). Ion pumps in the living cell. *Science J.* pp. 54–9.

Rossen, R., Kabat, H. & Anderson, J.P. (1943). Acute arrest of cerebral circulation in man. *Arch. Neurol. Psychiat.* (Chicago), **50**, 510–34.

Rostan, Léon. (1823). *Recherches sur la ramollissement du cerveau.* Paris.

Schneider, M. (1957). 'The metabolism of the brain in ischaemia and hypothermia' in *Metabolism of the nervous system.* Ed. D. Richter. London.

Schwann, Theodor (1847). *Microscopical advances into the accordance in the structure and growth of animals and plants.* Translated from the German by Henry Smith. Printed in London for the Sydenham Society.

Stensen, N. (1776). 'Discours sur l'anatomie du cerveau dans une assemblée chez M. Thevenot en 1668' in *Exposition anatomique sur la structure du corps humain* by J.-B. Winslow. Vol. IV, Paris. pp. 147–92.

Stirling, W. (1902). *Some apostles of physiology.* London. Printed privately by Waterlow and Sons.

Stokes, William (1854). Diseases of the heart and aorta.

Thudichum, L. J. W. (1863). *A treatise on gall-stones and their chemistry, pathology and treatment.* London.

Thudichum, L. J. W. (1883). On chemical diseases of the brain and spinal cord as conditioned by the chemical constitution of these organs. *Brit. Med. J.* **2**, 524–6.

Thudichum, L. J. W. (1884). *A treatise on the chemical constitution of the brain.* London.

Vauquelin, L. N. (1812). Analyse de la matière cérébrale de l'homme et de quelques animaux. *Ann Chim.* **81**, 37–74.

Vernon, H. M. (1911). Quantitative estimation of indophenol oxidase in animal tissues. *J. Physiol.* **42**, 402–27.

Vesalius, A. (1543). *De humani corporis fabrica libri septem.* Basel.

Waller, A. (1850). *Phil. Trans. Roy. Soc. Lond.* pp. 423–9.

Waller, A. (1852 a). Sur la reproduction des nerfs et sur la structure et fonctions des ganglions spinaux. *Müller's Archiv für Anat. Physiol., u. wissenschaftliche Medicin,* pp. 392–401.

Waller, A. (1852 b). Examens des altérations qui ont lieu dans les filets d'origine du nerf pneumogastrique et des nerfs rachidiens par la suite de la section de ces nerfs au-dessus de leur ganglions. *Compt. rend. hebd. Acad. Sci.* **34**, 842–7.

Waller, A. (1852 c). Septième mémoire sur le système nerveux. *Compt. rend. hebd. Acad. Sci.* **35**, 301–6.

Waller, A. (1852 d). Huitième mémoire sur le système nerveux. *Compt. rend. hebd. Acad. Sci.* **35**, 561–4.

Waller, A. (1857). Expériences sur la section des nerfs et les altérations qui en résultent. *Compt. rend. des séances et mémoires de la Soc. Biol.* **3** (2nd series), 6–8.

Warburg, O., Posener, K. & Negelein, E. (1924). Über den Stoffwechsel der Carcinomzelle. *Biochem. Z.* **152**, 309–44.

Weinberger, L., Gibbon, M. H. & Gibbon, J. H. (1940 a). Temporary arrest of circulation to the central nervous system: physiologic effects. *Arch. Neurol. Psychiat.* (Chicago), **43**, 615–34.

Weinberger, L., Gibbon, M. H. & Gibbon, J. H. (1940 b). Temporary arrest of circulation to the central nervous system: pathological effects. *Arch. Neurol. Psychiat.* (Chicago), **43**, 961–86.

Weiss, P. & Hiscoe, H. B. (1948). Experiments on the mechanism of nerve growth. *J. exper. Zool.* **107**, 315–95.

Willis, T. (1681). *The remaining medical works of that famous and renowned physician Dr Thomas Willis.* Vol. IV. Of the accension of the blood. Vol. VI. Of the anatomy of the brain. Translated by S. Pordage. London.

Winterstein, H. (1906). Zur Frage der Sauerstoffspeicherung. *Zentrallblatt für Physiologie.* **20**, 41–4.

Winterstein, H. (1907). Der respiratorische Gaswechsel des isolierten Froschrückenmarkes. *Zentrallblatt für Physiologie.* **21**, 869–73.

5

THE EVOLUTION OF IDEAS ABOUT ANIMAL HORMONES

by F. G. Young

THE ORIGIN OF THE IDEA OF HORMONES

The work of E. H. Starling

The *Oxford English Dictionary* (volume entitled 'Supplement and Bibliography' published by the Clarendon Press, Oxford, 1933) defines a hormone as 'a substance formed in one organ and carried by the blood stream to another organ which it stimulates'. The Dictionary ascribes the birth of the word to 1905 and to the published account of E. H. Starling's Croonian Lectures to the Royal College of Physicians of London (Starling, 1905). In these four lectures, published under the title 'The Chemical correlation of the functions of the body', Starling generalized the idea of a humoral control of physiological function which he had earlier developed with Bayliss. This idea had emerged from the results of their joint investigations of the control of the secretion of pancreatic juice, which resulted in their discovery of the substance that they named 'secretin' (Bayliss & Starling, 1902). In 1905 Starling wrote that 'chemical messengers'—or 'hormones' (from ὁρμάω I excite or arouse) as we might call them, have to be carried from the organ where they are produced to the organ which they affect by means of the blood stream, and the recurring physiological needs of the organism must determine their repeated production and circulation throughout the body' (1905, p. 340).

In their paper of 1902 Bayliss and Starling state that the independent researches of Popielski in 1900, and of Wertheimer and Lepage in 1901, had revealed that 'the introduction of acid into the duodenum still excites pancreatic secretion after section of both vagi and splanchnic nerves, or destruction of the spinal cord, or even after complete extirpation of the solar plexus. Popielski concludes, therefore, that the secretion is due to a peripheral reflex action' (Bayliss & Starling, 1902, p. 326). Bayliss and Starling started their joint investigation of the nature of this reflex by verifying the observations made by the earlier investigators. Like them, they failed to demonstrate experimentally that a nervous mechanism was indeed an essential factor in the reflex secretion of pancreatic juice; but an alternative explanation was not immediately obvious. Because the introduction of hydrochloric acid into the upper part of the small intestine elicited the secretion of pancreatic juice in the absence of all obvious nervous connexion between the intestine and the pancreas, they stripped off the mucous membrane from part of the small intestine, ground this mucosal material with sand and 0·4 per cent. HCl, and filtered the extract through cotton wool. They then injected an extract prepared in this way into the veins of a suitably prepared animal, and observed that a substantial stimulation of the secretion of pancreatic juice soon followed. They inferred from this exciting observation that the acid material which entered the duodenum from the stomach liberated from the duodenal mucosa a substance which then entered the blood stream and, reaching the pancreas, there stimulated the secretion of digestive juice by that organ. The name they coined for this substance was 'secretin'. Bayliss and Starling thought that the term 'internal secretion', a term commonly in use at that time to describe the liberation of material into the blood stream as opposed to its release into a special duct, did not emphasize sufficiently the biological significance of a chemical substance which was produced in one part of the body and carried in the blood to another, where it could induce a physiologically important change. They discussed possible neologisms for the general name for agents of this sort, and ultimately the term 'hormone' was suggested to Starling by William Bate Hardy, after Hardy had consulted W. T. Vesey, his classical colleague at Gonville and Caius College, Cambridge (see Rolleston, 1936, p. 2). The word 'hormone' was at first criticized because, by its derivation, it should apply only to those internal secretions which exert a stimulating action, whereas under some conditions certain internal secretions appear to exert a depressing action. But most investigators did not worry about any need to distinguish excitation and depression in this connexion. Both processes are concerned with the co-ordination of function.

Today one might conveniently define an animal hormone as 'a substance which is liberated into the blood stream or other circulating fluid and which catalyses changes in some or all tissues'. On the basis of such a definition there is difficulty in deciding whether some substances are hormones or not, but the definition seems to be sufficiently precise and to be of general value. The term 'endocrine organ' or 'endocrine tissue' is a convenient description of the tissue of origin of a 'hormone' or 'endocrine secretion', the latter terms being used synonymously.

The word 'hormone' has been applied to many different types of substance, including those involved in the development and growth of plant cells, the warning substances produced by minnows, the sexual attractants of moths, butterflies, cockroaches and crustaceans, and the so-called recognition odours. There has been much study of hormone action in insects. This essay will be concerned only with some of the hormones in higher mammals, including man.

The discovery of internal secretion

Sometimes the great French physiologist Claude Bernard is credited with the first formulation of the view that substances can be liberated by the tissues directly into the blood stream, and of first suggesting the existence of what we now consider to be a hormone. Certainly in 1853 Bernard published clear evidence that the liver liberates glucose into the blood stream and concluded (I quote a translation of this words) 'Sugar is manufactured in the liver which must therefore be considered as an organ which produces or secretes sugar' (Bernard, 1853, p. 54). Bernard developed this idea, and in 1859 he wrote 'organs which provide secretions that are purely internal are the spleen, the thyroid, the supra-renal capsules, etc.' (Bernard, 1859, vol. 2, p. 441). Glucose is certainly a substance liberated

into the blood stream in order to bring about a biological action
in the tissues of the body in general, but it is not now usually
regarded as a hormone. It does not primarily exert a catalytic
action in tissues but provides fuel for many tissues, and the
term 'hormone' is now restricted to substances whose function
is not mainly or primarily that of providing metabolic fuel.

The belief that what we now call endocrine organs might
secrete substances into the blood stream was in fact current
before Claude Bernard's writings about the liberation of sugar
into the blood stream. For example in the *Cyclopaedia of anatomy
and physiology*, edited by R. B. Todd between the years 1836 and
1859, there is an article in volume 4 (1852) by W. B. Carpenter
(pp. 439–72), which is entitled 'Secretion' and in which occurs
the following words.

We refer to that elaborating agency, which is now generally believed
to be exerted upon certain materials of the blood by the spleen,
thymus, and thyroid glands, and suprarenal capsules (which are
sometimes collectively termed vascular glands) [and which are]...
unprovided with excretory ducts for the discharge of the product of
their operation. These products, instead of being carried out of the
body, are destined to be restored to the circulating current, apparently in a state of more complete adaptiveness to the wants of the
nutritive function; in other words, these vascular glands are concerned in the assimilation of the materials that are destined to be
converted into organized tissues, instead of being the instrument
of the removal of the matters which result from the disintegration
or decay of those tissues (p. 440).

One can see from this quotation that a belief that the vascular
glands, or 'blood glands' as they were sometimes then called,
were sources of an internal secretion, was regarded as generally
accepted by the author of an article which appeared in the
early 1850s. Another article in the same *Cyclopaedia*, by
Heinrich Frey, was on the 'supra-renal capsules', which are now
generally known as the adrenal glands. The article, translated
from the German, contains the following statement

Their contents [that of the supra-renal capsules] are not exuded
outwardly, as are those of the gland-vesicles previously described,
but into the fibrous framework of the organ, in which they exist in

the fluid form, and from which they are subsequently received into the vascular system either by immediate or mediate resorption. We are therefore correct in regarding the supra-renal capsules as glandular organs, and their function as secretory (vol. 4, 1852, p. 839).

One can infer that in describing the liberation of glucose into the blood by the liver as an 'internal secretion' Claude Bernard was adopting for the liver a view which was already accepted about 'vascular glands' in general, and about the adrenal glands in particular.

But long before this there had been foreshadowings of the idea of a hormone or internal secretion. In 1670 Richard Lower, disagreeing with the ideas of Galen and Vesalius, who believed that the pituitary gland discharged pituitia (that is phlegm or mucus—the waste material from the brain) through the infundibulum into the nose, wrote 'whatever serum is separated into the ventricles of the brain and issues out of them through the infundibulum into the *glandula pituitaria* distills not upon the plate but is poured into the blood and mixed with it' (Lower, 1670). Lower, who was born in 1631, became assistant to the celebrated Thomas Willis at Oxford and there is some doubt whether Lower was himself the originator of the idea that the pituitary gland liberates a substance into the blood stream, or whether Willis was responsible for this pertinent suggestion.

The first formulation of the conception of what we now think of as a hormone has been ascribed to Théophile de Bordeu on the basis of his verbose work *Recherches sur les maladies chroniques* (1775). In this work the suggestion is made that each organ of the body gives off emanations which are necessary and useful to the body as a whole. The writer thus envisaged emanations not only from endocrine organs but also from parts of the body not usually included among the endocrine glands. But de Bordeu particularly laid stress upon the tonic effects of testicular and ovarian emanations, and their influence on the secondary sex characteristics as is demonstrated by the obvious alterations in the body in general which follow castration. Such changes have of course been long recognized on the farm and in the harem, but in de Bordeu's time the need for experimental investigation was not yet sufficiently appreciated to induce him to test his belief experimentally, and his view remained without experimental basis. Nevertheless, because of his potent speculations and perhaps of his charm, de Bordeu gained a fashionable practice at the Court of Louis XV. Before this, in 1766, Albrecht Haller had grouped together the thyroid gland, the thymus and the spleen, the three organs mentioned by W. B. Carpenter in 1852 in the work cited above, as glands without ducts which pour a special substance into the veins and so into the general circulation. But this and other similar guesses made during the eighteenth century stimulated no investigation of them, and these suggestions were almost completely sterile.

Early in the next century T. Wilkinson King of Guy's Hospital, London, in discussing the structure and function of the thyroid gland, once again in 1836 put forward the idea of an internal secretion. Moreover, around this time George Gulliver, by microscopic observation, had seen minute spheroidal bodies both in the substance of the adrenal glands and in their veins, and had concluded that the veins of the adrenal gland were the excretory ducts of this organ. In 1840 he wrote 'the adrenals pour into the blood peculiar matter which has doubtless a special use and is still an interesting and important subject for further enquiry'. Today this comment still has much truth in it.

A clear early experimental indication of the existence of hormones came from the observations of A. A. Berthold of Göttingen, Germany, who found in the middle of last century that transplantation of the testes of a cock to another part of the body of the same animal prevented the atrophy of the comb which normally follows castration, even though the bird remained sterile. In 1849 Berthold described the influence exerted by the testes on the blood and so on the body as a whole, but his important experimental demonstration of the existence of an internal secretion, and the establishment of an endocrine function of the testes, appears to have exerted no influence on contemporary scientific thought. His observation was indeed forgotten until the scientific communication in which it was embalmed was exhumed by Biedl in 1910.

Ideas about animal hormones

Thomas Addison

To the observations of Thomas Addison, of Guy's Hospital, London, is often ascribed the foundation of modern endocrinology. Almost simultaneously with the publication of Berthold's important but unrecognized discovery Thomas Addison was making observations about what is now known as 'Addison's disease', a condition which he observed to be associated with a diseased condition of the supra-renal capsules, or the adrenal glands as I shall usually describe them. The adrenal glands were known as the supra-renal capsules at one time because the interior part, or the medulla, easily undergoes liquefaction after death. There had long been discussion and argument as to whether or not the gland contained a cavity in the living animal and man. In 1611 Bartholinus first described them as *capsulae atrabilariae* or atrabilious capsules, and suggested that the black bile, which he believed them to contain, passed to the kidneys and thence to the urine. Cassan, in 1789, argued that, since the adrenal glands are larger in Negroes than in Europeans, they might be concerned in the formation of pigment in the skin, and there is a possibility that this unsubstantiated speculation may have aroused the interest of Thomas Addison in these glands, since he was a physician with a particular interest in dermatology and had early observed the pigmentation of the skin which is often seen in Addison's disease.

In a publication in 1849 Addison described what is now known as Addisonian anaemia, or pernicious anaemia, in some cases of which he had noticed, at post-mortem examination, a diseased condition of the adrenal glands. He inferred that the adrenal glands might be concerned in the production of blood. In this publication there is some confusion between pernicious anaemia and what is now known as Addison's disease, but in 1855 Addison published a monograph entitled 'On the constitutional and local effects of diseases of the supra-renal capsules' in which he wrote 'the leading and characteristic features of the morbid state to which I would direct attention are anaemia, general langour, and debility, remarkable feebleness of the heart's action, irritability of the stomach, and a peculiar change of the colour of the skin, occurring in connexion with a diseased condition of the supra-renal capsules'. He described 'a dingy or smoky appearance of the skin' which might also show 'various tints or shades of deep amber or chestnut brown' (Addison, 1855).

Addison's monograph of 1855 is mainly devoted to a description of eleven cases of the condition, most of which had been reported by others. Subsequent comment has thrown doubt on the validity of the inclusion of a number of them—especially those in which only one adrenal gland was diseased. Addison's own loyal friend and pupil, Sir Samuel Wilks, accepted only four of these eleven cases as truly Addison's disease. Nevertheless Addison's observation and intuition were not at fault in his main contention that a diseased condition of the adrenal glands is significantly associated with a distinctive syndrome, although one must agree that he proved nothing that would satisfy the critical comment of a modern statistician.

But it was not statistical analysis that made Addison's contemporaries and successors hesitate about the significance of Thomas Addison's observations. Addisonian anaemia and Addison's disease (an eponym first applied by Trousseau in 1856) had at first been confused together, and disentanglement of the characteristics of the two conditions took some time. Again the comparative rarity of Addison's disease was probably a reason why less attention was paid to Addison's description of it than might otherwise have been expected. Moreover the idea that a disease, and particularly a fatal one, could arise from a disorder of something as small as the adrenal glands ran counter to the quantitative mechanistic ideas that were fashionable at that time, no doubt as the result of the rapid and practically effective development of physics.

Addison appears first to have considered that the symptoms that he had observed arose from an interference with a secretory activity of the adrenal glands, even though, as he frankly admitted, the function of these glands was almost or entirely unknown. But the custom of the time was to explain the phenomena of health and disease largely in terms of nervous action and its disturbance. As the late Sir Frederick Gowland Hopkins wrote in 1938 'Up to near the end of the last century nearly every expert looked to the influence of the nervous system alone as concerned with the coordination of functions of the body; the conception of chemical regulation and co-ordination had achieved no place in the minds of the majority' (Hopkins, 1938). Addison himself was apparently uncertain about the interpretation of his observations, and is reported by Rolleston to have said in a discussion at the Royal Medical Chirurgical Society in 1858:

We know that these organs [the adrenal glands] are situated in the direct vicinity and in contact with the solar plexus and semilunar ganglia, and receive from them a large supply of nerves, and who can tell what influence the contact of these diseased organs might have on these great nerve centres and what share that secondary effect might have on the general health and in the production of the symptoms presented? (1936, p. 340).

Addison was not alone in considering that the condition he had described might arise from an influence on local nerve centres of the disease of the adrenal glands, and in 1882 J. F. Goodhart stated that Addison's disease is not due to any special change in the adrenal glands but to chronic inflammation spreading to and strangulating the abdominal sympathetic nerves (Goodhart, 1882).

Thomas Addison, like Claude Bernard, was a somewhat lonely person. He was subject to fits of depression and, in 1860, ended his life by suicide at Brighton, shortly after he had retired from his post at Guy's Hospital.

EARLY EXPERIMENTAL INVESTIGATION OF THE
FUNCTION OF ENDOCRINE GLANDS

The effects of removal of the adrenal glands

An important effect of the publication of Addison's monograph in 1855 was the stimulation of direct experimental research into the function of the adrenal glands. Vulpian stated in

1856 that he had observed that both the medulla of the adrenal glands and the venous blood from these organs stains green with iron salts. He believed that this observation must be significant with respect to the function of the glands. Unfortunately, but perhaps understandably, the ebullient and somewhat unreliable French–American Brown-Séquard was the first to investigate the effects of the experimental removal of the adrenal glands. Brown-Séquard (1856) found that excision of both adrenal glands from animals was invariably and rapidly fatal, and concluded that the adrenal glands were indispensible for life. Nevertheless in the following year Phillipeaux stated that rats could survive the removal of both adrenal glands and explained the fatalities observed by Brown-Séquard in his experiments as a result of damage to other tissues, and in particular to the nervous system. The results of the later investigations of others agreed with Phillipeaux's observations. Brown-Séquard's fatalities probably resulted not from the removal of the adrenal glands but from sepsis and traumatic shock, the avoidance of which would have prevented the death of his animals; his deductions were not justified by the experimental evidence he provided at that time. Nevertheless Brown-Séquard later expanded his views on internal secretions, and in 1889 he described his own rejuvenation as a result of the self-administration of an extract of testis tissue. He thus widely publicized the idea of the potency of internal secretions, although, as Herbert Evans has picturesquely written, in this way 'Endocrinology—suffered obstetric deformation at its very birth' (1933).

Adrenaline and the adrenal glands

Sir Henry Dale has entertainingly related the traditionally recorded account of the discovery of adrenaline (Dale, 1948), although this has been slightly modified by Barcroft & Talbot (1968) on the basis of earlier written accounts. Dr George Oliver was a physician of Harrogate who employed his winter leisure hours in experiments on human subjects with the aid of apparatus of his own devising. One of his instruments was designed to measure the calibre of peripheral arteries, and Dr

Ideas about animal hormones

Oliver measured the diameter of the radial artery before and after he had administered to his son an extract of adrenal glands prepared from material supplied by the local butcher. Oliver thought that he could detect a constriction of the artery as the result of the administration of the extract by mouth and went to London to tell Professor E. A. Schäfer, Professor of Physiology at University College London, what he believed he had observed. He found Professor Schäfer engaged in an experiment in which the blood pressure of a dog was being recorded and Schäfer was, not unnaturally, sceptical about Oliver's story and perhaps impatient at the interruption. But Oliver was in no hurry and, producing a vial of his adrenal extract from his pocket, asked only that some of it should be injected into the vein of the dog when Professor Schäfer's own experiment was finished. And so Schäfer gave the injection and then stood amazed to see the mercury mounting in the arterial manometer until the recording float was almost lifted out of the instrument. In this way was discovered the extremely active pressor substance adrenaline (sometimes known as epinephrine), which is present in the adrenal medulla, the inner part of the gland that easily liquefies after death (Oliver & Schäfer, 1894). In further experiments Oliver and Schäfer collaborated in tests of the effect on blood pressure of extracts of a number of tissues whose function was then unknown. In this way they discovered the presence of vasopressin in the pituitary gland, this substance ultimately being located in the posterior portion of that gland.

In 1901 adrenaline was isolated from adrenal glands both by Aldrich and by Abel in the United States, and shortly before this in Japan by Takamine, who had earlier worked in Abel's laboratory. The final step in the isolation involved the addition of strong ammonia to an extract and this addition produced a crop of crystals of adrenaline. But, as long afterwards became clear, the material prepared in this way contained the adrenaline but not the noradrenaline which was also present in the adrenal glands and which was therefore missed for many years.

Adrenaline was the first hormone to be isolated in a crystal-

135

line form and the first to be chemically synthesized, the synthesis being achieved by Stolz in 1904, and independently by Dakin in 1905. Disappointingly, this pharmacologically powerful substance was found to be ineffective in the treatment of Addison's disease, and the latter condition was assumed to be due to a deficiency of something produced by the outer part of the adrenal glands, the adrenal cortex. Moreover, experimental evidence showed that removal of the adrenal cortices led to death, while excision of the adrenal medullae, which contained adrenaline, did not. So adrenaline from the adrenal glands was apparently not essential for life.

In 1904 the correspondence between the effects produced by the injection of adrenaline and those produced by stimulation of the nerves of the sympathetic nervous system led to the prescient suggestion by T. R. Elliott that the liberation of adrenaline at sympathetic nerve endings, as the result of the arrival of an impulse down the nerve, might transmit the excitatory or inhibitory action of the impulse to the effector cells, muscle or gland. In this way began what might be described as a hormonal attack upon the central nervous system itself.

Thyroid gland

Perhaps as the result of the publication of the results of Brown-Séquard's experiments on the removal of the adrenal glands from animals Schiff, in 1858, described the effects of removal of the thyroid gland from animals, finding that this operation was fatal. Enlargement of the thyroid gland, or goitre, had long been recognized as a condition not uncommon in people living far from the sea, and stunting and mental deficiency in infancy, described as cretinism, had been associated with the existence of endemic goitre, this association having been described by Paracelsus in the sixteenth century. When, in 1812, Courtois isolated iodine from the ash of sponges, Coindet was led to investigate the use of iodine in the treatment of goitre, and for a time the administration of iodine appeared to be useful in this respect. But treatment with iodine could be dangerous and the therapeutic use of this substance was abandoned for a while. Sir William Gull, of Guy's Hospital

London, in 1875 described a cretinoid condition in adult women and associated it with atrophy of the thyroid gland, and a few years later Ord dubbed this condition 'myxoedema'. In 1884 Schiff described a continuation of his earlier investigation (of 1858) about the effects of the experimental removal of the thyroid gland, and in the same year Horsley, on the basis of similar investigations, supported the view that myxoedema, cretinism and the often fatal effects of experimental removal of the thyroid gland, were all due to the loss of a supposed internal secretion of this organ. In 1891 G. R. Murray successfully treated patients with myxoedema by the injection of a simple extract of thyroid tissue, and subsequently observed that even simpler treatment, involving the taking of thyroid tissue by mouth, was equally effective.

In the 1890s successful treatment by iodine of some instances of goitre was rediscovered, and on general grounds the belief became widespread that the thyroid gland, and perhaps its supposed internal secretion, must contain the element iodine. But for long the chemical methods then available were not sufficiently delicate for a demonstration of the existence of iodine in the thyroid gland to be convincing. Ultimately, in 1896, Baumann showed clearly that thyroid tissue does indeed contain iodine, and that the amount varies, usually being in the range 0·05–0·45 per cent.

The pancreas and diabetes mellitus

The existence of diabetes mellitus has been known for at least two thousand years, but until 1889 there was little evidence about the cause of the condition. In that year von Mering & Minkowski (1889–90) found that a condition which resembled human diabetes mellitus followed surgical removal of the pancreas from a dog. At the time of this discovery Minkowski, who was born in 1858 and died in 1931, was an assistant in the Medical Clinic of the University of Strasbourg, under the direction of Naunyn. There have been some doubts as to how this discovery came to be made. But a letter written by Minkowski himself in 1926 has been published (Houssay, 1952), which gives an account that is probably the most authentic available. In April 1889 von Mering was working in Hoppe-Seyler's Institute at Strasbourg, while Minkowski was in the Medical Centre. Going over to Hoppe-Seyler's Institute to consult in the library some chemical periodicals which were not available in the Clinic, Minkowski casually met von Mering and a discussion about lipanin began. Von Mering had shortly before recommended lipanin, an oil with 6 per cent. free fatty acid in it, for the treatment of various digestive disturbances in the belief that its apparently favourable effect might be attributable to its free fatty acid content. Minkowski rather scoffed at the use of lipanin, and as a result some argument developed as to whether the pancreas really performs an important function in assisting the splitting of fatty acids in the gut. As the result of this disagreement the two decided to investigate together the effects on the digestion of fat and on health in general, of the surgically difficult operation of removal of the pancreas from a dog, the dog being supplied by von Mering who also gave assistance with the operation. The day after the operation had been completed von Mering had to go away for a time because of family illness and the dog was kept by Minkowski to await von Mering's return in order that the proposed experiments on the digestion of fat could be carried out. Because there was no suitable cage available the dog was kept tied up in the laboratory, and although the animal had been house-trained and was frequently taken out by a laboratory assistant, it even more frequently passed water on the floor of the laboratory. According to Minkowski, he had been taught by Naunyn always to test for the presence of sugar in the urine if polyuria was observed, so he tested the urine of the dog for the presence of sugar, finding there to be present something like 12 per cent. The animal appeared to be suffering from a form of diabetes mellitus. The story which is often repeated that glycosuria was first suspected because a laboratory assistant noticed that flies, attracted by the sweetness, settled wherever the dog had passed urine, has been directly contradicted by Minkowski (see Allen, Stillman & Fitz, 1919, revision of footnote 37, p. 38), but such an observation could have been made independently.

Ideas about animal hormones

According to Minkowski the pursuit of research on the experimental condition of diabetes mellitus which had resulted from removal of the pancreas was a matter which did not greatly interest von Mering, and although von Mering assisted with the removal of the pancreas from a few more animals, and also carried out some determinations of glycogen on liver tissue, he was not otherwise concerned in the development of the investigation. The publication in which the results of this tremendously important observation appeared was described as an investigation by von Mering and Minkowski because von Mering came before Minkowski in alphabetical order and also because von Mering was substantially the senior of the two. The inference has subsequently incorrectly been drawn that von Mering's contribution was the major one in this important joint discovery.

The investigations carried out after publication of the first paper were entirely by Minkowski and others, who observed that experimental diabetes mellitus develops in many different species of carnivorous animal after surgical removal of the pancreas. Minkowski also attempted to prepare extracts of the pancreas the administration of which would relieve the experimentally-produced diabetic condition, but in this, like many others, he was unsuccessful. In 1901 Opie (1900–1) described damage to the islets of Langerhans of the pancreas in patients with naturally occurring diabetes mellitus, and this observation strengthened the view that diabetes mellitus, both natural and experimental, resulted from the loss of something produced by the pancreas, probably by the islets of Langerhans which are embedded in that organ. But attempts to prepare the hypothetical hormone were unsuccessful for many years. Only with the investigations of Banting and Best, published in 1922, was insulin prepared and made available for the therapeutic use which is widespread today.

But insulin was long in coming, and in the climate of opinion last century the criticism was natural that removal of the pancreas itself was not the cause of the experimentally induced diabetic condition, and that this condition resulted from damage to adjacent nerve centres. This criticism was effectively answered by Hédon (1893), who showed that if the major part of the pancreas was removed but a small piece of the tissue transplanted, with its circulation intact, to the surface of the abdomen, diabetes did not develop. But a severely diabetic condition at once appeared if the pancreatic graft was excised in circumstances in which the abdomen was not opened. Moreover, since there was no nerve supply to the graft, the control of secretion of the assumed hormone (insulin) appeared not to depend upon nervous stimuli.

DEFICIENCY AND EXCESS IN THE SECRETION OF HORMONES

During the latter part of last century, and in the early part of the present one, the view became accepted that an insufficiency in the amount of the internal secretion liberated by an endocrine gland into the blood could be a cause of disease. Furthermore the symptoms of such a disease might be curable by the repeated administration of extracts of healthy specimens of the gland in question. The converse then naturally had to be considered, namely that excessive activity of an endocrine gland might be the cause of another type of disease, the treatment of which might require partial or complete removal of the gland in question. The realization that exophthalmic goitre or Graves' disease results from overactivity of the thyroid gland was an early fruitful outcome of such a line of thought. As the number of recognized hormones grew the number of diseases which could be ascribed to the secretion of an excessive amount of a hormone rose *pari passu*. Hyperthyroidism, hyper-adrenalism, hyperpituitary diseases and so forth became more easily identifiable once the relevant hormones were available for investigation of their pathological as well as of their physiological actions.

HORMONES IN THE TWENTIETH CENTURY

By the early years of the present century, the existence of a number of mammalian hormones was surmised but only a few of them had been characterized. As we have already seen, adrenaline was the first to be obtained in pure form and chemically synthesized. Although secretin was of outstanding significance in the evolution of thought about hormones, Bayliss and Starling did not succeed in isolating this substance, and we now know that they could not have hoped to succeed at that time, since secretin is a polypeptide of moderate molecular weight and the isolation of such polypeptides has been successfully accomplished only in recent times.

Thyroxine, the iodine-containing hormone of the thyroid gland, was obtained in a crystalline form for the first time on Christmas Day 1914 by E. C. Kendall. Its structure was elucidated by C. R. Harington in 1926 and its chemical synthesis achieved by Harington in collaboration with Barger in 1926–7.

Although at the end of the second decade of the present century there was evidence that the islets of Langerhans of the pancreas, the gonads, and the adrenal cortex all secrete hormones of their own, the isolation of these then hypothetical substances had not been achieved. In a like position were the parathyroid glands, and the various parts of the pituitary gland or hypophysis. Investigation of the pituitary gland was hampered by the almost impregnable position in which nature had placed this tiny but important organ.

After the 1914–18 war the pancreas was the first endocrine organ to yield physiologically significant preparations of a hormone. The work of the Canadians Banting and Best in this connexion is well known but a brief account of the story must be included for the sake of completeness.

The isolation of insulin

In 1921 Banting was a young surgeon engaged in part-time teaching of physiology at the University of Western Ontario, London, Ontario. In the course of the preparation of a lecture on the pancreas he had the idea, while in bed late at night, that failure to extract from pancreatic tissue active preparations of its assumed anti-diabetic hormone was due to destruction of the hormone by digestive enzymes of the pancreas which were extracted along with it.

F. G. Young

At that time there was evidence that the putative hormone of the pancreas was produced in the small islands of special tissue scattered throughout the bulk of the organ. These had been named 'islets of Langerhans' after Paul Langerhans, who had first observed them in 1869. So confident was de Meyer (1909) that the islets of Langerhans secreted a hormone a deficiency of which could result in diabetes mellitus that in 1909 he named the then unborn hormone 'insuline'. If the pancreatic duct, through which pancreatic juice is secreted, is tied, the enzyme-secreting tissue of the pancreas, the acinar tissue, rapidly degenerates but the tissue of the islets of Langerhans does not. Banting had the idea that an extract of a pancreas in which this degeneration of the enzyme-containing tissue had previously been induced might yield the undamaged hormone of the pancreatic islets by a simple extraction procedure. Although he did not know it, Banting's idea was not new. It had provided the basis for an apparently unsuccessful investigation by Gley among others, many years before. But Banting, with the ignorance and enthusiasm of youth, decided in the spring of 1921 that he must find a means to investigate his idea experimentally. The Professor of Physiology at the University of Western Ontario suggested that he should go to see Professor J. J. R. Macleod, Professor of Physiology at Toronto, who had carried out much research on diabetes mellitus. Macleod received Banting in a kindly, sceptical manner, and pointed out that previous investigators had already had Banting's idea, but had not succeeded in preparing the supposed hormone. Why did Banting believe that he could succeed where others had failed? Banting had no complete answer to this question except that he was determined to isolate the hormone if it were at all possible. Macleod, who liked to allow a young man with an idea to try it out if possible, agreed that during the summer vacation of 1921 Banting might come to Toronto and carry out a limited number of experiments on dogs to see whether his idea would work in practice. Furthermore, Macleod said that he would make available as an assistant for Banting one of the medical students who wished to do research, during the vacation, before proceeding to clinical studies. And so Banting and Best began their investigation in the hot summer of 1921.

Their success in the preparation of insulin can I believe be ascribed in large degree to certain personal qualities of the two investigators—the obstinacy and determination of Banting and the analytical and experimental skill of Best. But another factor of some significance was the recent availability of methods which made possible the measurement of the sugar in a relative small amount of blood. In Claude Bernard's investigations 25 cc. of blood were needed for the estimation of its sugar content, and many estimations were obviously not possible on an experimental animal. But the development of micro-methods for the estimation of the constituents of blood, stimulated by Bang in 1913, made feasible the frequently repeated determination of blood sugar. Banting & Best (1922) at first used the method of Myers & Bailey (1916) for the estimation of blood sugar in their investigation, but later turned to the then recently published method of Shaffer & Hartmann (1921). In a personal communication Professor C. H. Best has written 'We needed 1 cc. for our blood sugar determinations but usually took about 3 cc. When we turned to the Shaffer–Hartmann method we could use 0·5 cc. as I remember it'. The administration of insulin in excess can quickly reduce the amount of sugar in the blood from the high values characteristic of diabetes mellitus to such a low one that the dangerous symptoms of hypoglycaemia appear. Some clinicians, for example Zuelzer (1908), may well have succeeded in preparing an insulin-containing extract of pancreatic tissue. But patients treated with such material were liable to exhibit alarming symptoms, which appeared to result from a generally toxic affect of the extract but which may in reality have resulted from hypoglycaemia induced by an overdose of insulin. In the absence of multiple estimates of the amount of sugar in the blood an investigator could be completely in the dark about the cause of such symptoms. These considerations emphasize both the importance of the use of animals in the early stages of a therapeutic investigation and the immense value to Banting and Best of continual estimates of the amount of sugar in the blood of their depancreatized dogs when these were treated with a simple extract of a pancreas in which the acinar tissue had been induced to degenerate by ligation of the pancreatic duct six weeks previously.

The publication of the discovery of insulin (Banting & Best, 1922) was quickly followed by the establishment of the efficacy of the treatment of human diabetes mellitus by repeated injections of insulin under controlled conditions. The industrial production of this hormone developed in an amazingly short time.

Insulin was obtained as a crystalline protein by Abel in 1926. The primary chemical structure of this hormone was elucidated by Sanger (see Sanger, 1960) and insulin was chemically synthesized by Zahn, by Katsoyannis and by Wang, with their respective collaborators, in the early 1960s.

The water solubility of hormones

The principle could be argued that since a hormone is liberated into the blood stream it must be soluble in aqueous media and that material obtained from a gland by extraction with aqueous solvents should contain its hormone. Banting and Best's original preparation of insulin was indeed a simple aqueous extract of the acinar-degenerate tissue from the duct-tied pancreas of a dog. Shortly after the discovery of Banting and Best, Herbert M. Evans in California prepared a simple aqueous extract of the anterior part of the pituitary gland and observed that its repeated administration could stimulate growth and influence the gonads of normal experimental animals. The idea that hormones might be obtainable by extraction of tissues with aqueous solvents also gained some support by the discovery of Aschheim and Zondek in 1926 that the urine of pregnant women contained a substance with the expected physiological actions of the supposed hormone of the ovary; that is to say the repeated administration of an extract of pregnancy urine to an animal from which the ovaries had been removed could prevent the changes in the secondary sex characters which normally follow castration of the female. Subsequently Koch and others showed that in male urine there existed a substance with the expected properties of the hormone of the testes, the repeated administration of this substance being able to prevent or cure the atrophy of secondary sex characters normally seen in the castrated male animal. In 1929 and 1930 Doisy in America, Marrian in Great Britain, and Butenandt in Germany, isolated from the urine of pregnant women two substances with ovarian hormone activity; subsequently Butenandt showed that these substances were related chemically to the sterols, of which there are large amounts in the animal body and in plants. Butenandt also isolated from normal male urine a substance with the capacity to reverse the changes in the secondary sex characters which follow castration in a male animal, and showed that this male hormone substance, androsterone, was also chemically related to the sterols. Ruzicka in Switzerland in 1934 converted cholesterol, a commonly occurring animal sterol, by purely chemical means, into androsterone. Later researches showed that the biologically active substances extracted from urine were metabolic products of the hormones secreted by the ovary and the testis respectively; but the differences in chemical structure and in physiological action, although significant, were not very great. Both the hormones and their metabolic products, though slightly soluble in water, were found to dissolve more easily in fat solvents.

The hormones of the adrenal cortex

Until the end of the 1920s attempts at extraction of the supposed hormone of the adrenal cortex were generally unsuccessful. Although in 1927 Rogoff and Stewart, and a little later Hartman and his colleagues, described the preparation of aqueous extracts of the adrenal glands the repeated administration of which could prolong the life of animals from which the adrenal glands had been removed, the activity of such extracts was very slight and the results uncertain. An entirely different method of attack was opened by Swingle and Pfiffner in 1930. These investigators extracted adrenal glands with ethanol, and this ethanolic extract in turn with benzene. The product was dissolved in aqueous medium for biological testing by injection into a suitable animal from which the adrenal glands had been experimentally removed. The activity of an adrenal extract prepared in this way in prolonging the life of adrenalectomized animals, and of patients with Addison's disease, was quickly and independently confirmed, and extraction of the tissue of the adrenal glands with fat solvents led to the subsequent developments in this field.

Why did Swingle and Pfiffner (1930) succeed where so many others had previously failed? I believe that the answer to this question comes under four main headings: (a) their facilities for biological tests were extensive. They used over 350 cats in their preliminary experiments; (b) they realized that aqueous extracts of adrenal cortical tissue, even if the cortex were dissected as free as possible from the medulla, were bound to contain adrenaline, or degradation products of adrenaline, all of which might exert an undesirable effect on test animals to which the extract was administered; (c) they appreciated the fact that treatment of adrenal tissue with a fat solvent might be more effective in extracting the active material than extraction with an aqueous medium. Since the fact that the blood contains substantial amounts of fatty material, including cholesterol, had been known for many years, there was no real basis for the view that a hormone secreted into the blood could not itself be a fatty substance; (d) finally, they were not afraid to administer what in those days appeared to be unreasonably large doses of extract. Before this, the assumption had been implicitly made (and it was an unjustified assumption) that if a gland secreted a hormone the gland would necessarily contain a relatively large amount of that hormone. But, as was subsequently calculated, the normal rate of secretion of the adrenal gland of an animal can exhaust the stored hormones in 6–12 seconds. The adrenal cortex in fact does not store a large amount of its hormones but manufactures them from acetate and cholesterol.

I may mention in passing that Swingle and Pfiffner showed that adrenaline and its basic degradation products could be removed from their extracts by filtration through permutite, an early application of ion exchange materials to biological investigation.

Ideas about animal hormones

Swingle and Pfiffner's researches were followed by the chemical investigations of Reichstein in Switzerland, and by Kendall and by Wintersteiner in the United States, and, during the 1930s altogether seven crystalline steroid substances were isolated from adrenal tissue, the repeated administration of any of which was capable of maintaining an adrenalectomized animal in health. In 1952 an eighth substance was obtained, aldosterone, as the result of the investigation of Simpson and Tait in Great Britain, its structure being determined and its chemical synthesis effected, in collaboration with Reichstein, Wettstein and their colleagues in Basle, Switzerland in 1954. No single substance is the hormone of the adrenal cortex, and there is little doubt that varying proportions of a number of the adrenal steroids can be liberated into the circulation under different conditions.

The pituitary gland

As we have seen, the protection which the pituitary gland enjoys, situated as it is at the base of the brain and inside the skull, delayed experimental investigation of the function of this complex organ. Anatomically, embryologically and functionally, there are three main divisions of the pituitary gland, although in the human being the whole complex organ weighs little more than one gramme. The *pars glandularis (pars distalis)* or anterior lobe is important in the control of metabolic functions in general. The *pars nervosa (processus infundibularis)* or posterior lobe, is concerned with the control of water balance and kidney function and can also affect the blood pressure (as first observed by Oliver and Schäfer), while extracts of the *pars intermedia*, or intermediate lobe, cause light frogs to darken by inducing dispersion of black pigment in their melanophores. The intermediate-lobe hormone can also stimulate the deposition of black pigment in the skin of mammals.

In 1884 Loeb observed that diabetes mellitus frequently occurs in patients in whom there exists a tumour of the pituitary gland, while Marie described the condition of acromegaly in 1886, a disease in which there is overgrowth of bone, and of the bones of the face and hands and feet in particular. But at this time Marie did not relate this disease to the pituitary gland. In 1887 Marie, and almost simultaneously but independently, Minkowski, stated that acromegaly was associated with, and probably due to, changes in the pituitary gland. The investigations of Oliver and Schäfer concerning the blood-pressure-raising substance in posterior pituitary tissue have already been described. Early experimental investigations suggested that surgical removal of the whole pituitary gland induces no obvious changes in an animal. These failures to induce a condition of pituitary deficiency probably resulted from incomplete removal of the anterior lobe of the gland, since removal of all but a small fragment of the anterior lobe can fail to result in the obvious effects of excision of the whole of the gland.

Although in 1912 Cushing (1912) and Aschner (1912) said that animals from which the pituitary gland had been removed no longer grew and suggested that acromegaly resulted from an excess secretion of a growth-stimulating hormone of the gland, these ideas were not unreservedly accepted at that time. In the 1920s Philip E. Smith at Berkeley, California, devised a method for the complete removal of the pituitary gland from the rat. He showed that excision of the whole gland was not fatal but that it resulted in atrophy of certain other endocrine glands, and in particular of the thyroid, of the adrenal cortex, and of the gonads, while general growth in young animals completely stopped. Smith furthermore showed that the continued administration of rat pituitary tissue to rats from which the pituitary had been removed could prevent or cure nearly all the symptoms which followed removal of the gland. Moreover, he was able to demonstrate that the anterior part of the pituitary gland was alone effective in this respect. In 1922 H. M. Evans and J. A. Long observed that the administration of a crude extract of ox anterior pituitary tissue to a normal rat increased its growth rate, and also brought about changes in the sex cycle and stimulation of the ovaries. Four years later Aschheim and Zondek discovered the presence in the urine of pregnant women of a substance which, like an extract of the anterior pituitary lobe, could stimulate the ovaries of an experimental animal. The presence of this pituitary-like substance in pregnancy urine was made the basis of a widely-used test for pregnancy.

Subsequently, as the result of the researches of Li and others, six different protein hormones have been isolated from anterior pituitary tissue. Four of these stimulate other endocrine organs to activity, namely corticotropin, thyrotropin and the two gonadotropins; in addition there is a growth-stimulating hormone and a hormone which can stimulate the secretion of milk under some conditions and which is known as prolactin. From the posterior pituitary lobe two peptide hormones have been isolated, vasopressin which can induce a rise of blood pressure and diminish the secretion of urine, and oxytocin which induces contraction of uterine muscle. The chemical structures of these substances have been completely elucidated and their chemical synthesis effected by du Vigneaud. The intermediate lobe of the pituitary gland yields two substances having a similar biological action, namely that of causing the deposition of melanin pigment in the skin of mammals and of inducing the expansion of the pigment-containing cells in certain amphibia in such a way as to induce darkening of the skin. These two forms of the melanocyte-stimulating hormone, as it is called, are chemically related to corticotropin from the anterior pituitary lobe. This fact may well be of significance in the darkening of the skin in Addison's disease which was described by Thomas Addison in 1855.

Houssay and his colleagues in Argentina found in the early 1930s that extracts of the anterior pituitary gland exert an antagonistic effect on the action of insulin and therefore an important controlling influence on carbohydrate metabolism. Young later observed that under suitable conditions the administration of anterior pituitary growth hormone can produce experimental diabetes in an animal, and can induce damage to the insulin-secreting cells of the pancreas of the islets of Langerhans so that a persistently diabetic condition develops. C. N. H. Long and Lukens found that antagonism also exists between the action of insulin and that of the secretions of the adrenal cortex. The discovery that hormones can act antagonistically to each other has been a most fruitful one. The important and powerful hypoglycaemia action of insulin is now realized to be kept under control in the body by the balanced antagonistic action of a number of other hormones.

CO-ORDINATION OF BIOLOGICAL FUNCTION AND DISEASE

Early last century researches on the functions of nerves by Magendie and Flourens in France and by Bell in Great Britain, placed the nervous system in a dominating position in ideas about biological co-ordination and control. Later Carl Ludwig and Pflüger in Germany and Claude Bernard in France, added their support to this view. In his researches concerning the liberation of glucose into the circulation by the liver Claude Bernard himself was at one time misled into believing that a nervous reflex was of importance in the control of this process (see Young, 1957). Perhaps last century the development of the electric telegraph provided a model for the idea that the nervous system is important in the co-ordination of function in the animal body. In general, telegrams are more exciting than letters and a co-ordination of function by means of hormones could have seemed in some ways analogous to the transmission of orders by the casting of letters into rivers. Moreover the domination last century of quantitative ideas in scientific thought may well have militated against acceptance of the view that the continual liberation into the bloodstream of minute amounts of hormones by very small organs could be a condition for the normal health of the whole body of a man. But by the turn of the century such an idea was no longer regarded as likely to be unsound. The view was then steadily gaining ground that ineffective functioning or experimental removal of an endocrine organ could cause the appearance of disease, and that this result could be prevented or cured by the continual administration of the hormone which was normally secreted by the gland in question. About the same time the corollary of this view began to be accepted, namely that disease could also arise from naturally-occurring malfunction of an endocrine gland which caused the production of an excess of its hormone—a biological excess which might be quantitatively minute. The availability of hormone-containing extracts of endocrine glands made feasible the experimental investigation of this possibility. And so the cautious application of surgery to the treatment of certain diseases of the thyroid gland and, much later, of the pituitary gland and of the adrenal glands, began to be explored.

The realization that in some instances hormones might act antagonistically to each other provided an added stimulus to an investigation of the possible therapeutic advantages of removal of endocrine tissues in disease, or the use of drugs which diminish the rate of secretion of hormones; these ideas led to, among other developments, the treatment of malignant complications of diabetes mellitus by removal of the pituitary gland.

THE CONTROL OF HORMONE SECRETION

The rate of secretion of the hormones of the adrenal cortex, of the gonads, and of the thyroid gland, is greatly influenced by the activity of the anterior pituitary gland in secreting its 'tropic' hormones, corticotropin, the two gonadotropins, and thyrotropin. The activity of the pituitary gland in secreting its hormones is influenced by nervous centres in the hypothalamus, the part of the brain immediately above the pituitary gland. We thus have an example of nervous co-ordination of function which is mediated by the secretion of hormones.

The brain influences the liberation of anterior pituitary hormones into the circulation through the mediation of 'releasing factors' originating in centres in the hypothalamus of the brain. In some instances there is evidence of what can be called a 'feed-back' control of secretion of some of these hormones whereby, for instance, the secretion of the adrenal cortex damps down the secretion of corticotropin by the anterior pituitary gland, perhaps by acting on the hypothalamus. There is also evidence that the secretion of one endocrine gland may act upon another gland to increase or to diminish the supply of the hormone of the second gland, and indeed a circle of action and reaction that is perhaps not vicious but is certainly difficult to elucidate, exists between some of the endocrine organs.

THE MECHANISM OF ACTION OF HORMONES

Hormones must directly or indirectly influence enzyme reactions, which themselves lie at the basis of metabolic changes. An influence of hormones on membrane permeability may be of fundamental importance in this respect.

The substance that is present in the endocrine gland is not necessarily the form in which a hormone circulates in the blood. Nor is the hormone in the blood necessarily identical with that substance which ultimately brings about a physiological effect in the tissues. The hormones present in the blood may themselves undergo metabolic changes in the blood or in the tissues before they are able to bring about a reaction on the cells and enzymes which they influence. So the growth of knowledge has meant that the precise definition of a hormone could become a matter of pedantic difficulty.

CONCLUSION

Chemically the mammalian hormones are a miscellaneous lot. The list includes proteins, peptides, amino-acids, amines, and steroids of various sorts, though proteins and peptides predominate. There is no obvious reason to expect that hormones should be substances of a very special sort elaborated by only one type of tissue. On general grounds one might suppose that they are metabolites to which responsiveness in tissues has developed during the process of evolution.

The idea that all the tissues of a complex body can produce minute amounts of all hormones is a general one that is hard to prove or disprove. One can say that in many instances endocrine glands are derived from tissues which, at one stage in evolutionary and embryological development, constituted a part of a barrier tissue between the external and the internal

Ideas about animal hormones

environments of an animal. Many of the hormones are themselves concerned with the movement of materials across barriers (Young, 1959, 1964) and may constitute part of a mechanism which damps down any tendency to excessive alterations in the composition of that internal medium in which the cells and intracellular structures of a complex organism exist.

REFERENCES

Addison, T. (1855). *On the constitutional and local effects of disease of the supra-renal capsules.* S. Highley, London.

Allen, F. M., Stillman, E. & Fitz, R. (1919). *Monographs of the Rockefeller Institute for Medical Research No. 11.* 'Total dietary regulation in the treatment of diabetes.'

Aschner, B. (1912). *Arch. ges. Physiol.* **146**, 1–146. 'Ueber die Funktion der Hypophyse.'

Banting, F. G. & Best, C. H. (1922). *J. lab. clin. Med.* **7**, 251–66. 'The internal secretion of the pancreas.'

Barcroft, H. & Talbot, J. T. (1968). *Postgrad. med. J.* **44**, 6–8. 'Oliver and Schäfer's discovery of the cardiovascular action of suprarenal extract.'

Bayliss, W. M. & Starling, E. H. (1902). *J. Physiol.* **28**, 325–53. 'The mechanism of pancreatic secretion.'

Bernard, C. (1853). *Nouvelle fonction du foie.* Baillière, Paris.

Bernard, C. (1859). *Leçons sur les propriétés physiologiques et les altérations pathologiques des liquides de l'organisme.* 2 Vol. Baillière, Paris.

Bordeu, T. de (1775). *Recherches sur les maladies chroniques.* Quoted by Rolleston, H. D. (1935), p. 16.

Brown-Séquard, C. E. (1856). *C. R. Acad. Sci., Paris.* **43**, 422–5. 'Recherches experimentales sur la physiologie et la pathologie des Capsules Surrénales'; 542–6. 'Recherches experimentales sur la physiologie des capsules surrénales.'

Cushing, H. (1912). *The pituitary body and its disorders.* J. B. Lippincott Co., Philadelphia.

Dale, H. (1948). *Brit. med. J.* **2**, 451–5. 'Accident and opportunism in medical research.'

Evans, H. M. (1933). *J. Amer. Med. Assoc.* **101**, 425–32. 'Present position of our knowledge of anterior pituitary function.'

Goodhart, J. F. (1882). *Trans. Path. Soc. Lond.* **33**, 340–5. 'Simple atrophy of the supra-renal capsules, accompanied by melasma

F. G. Young

supra-renale and other symptoms of Addison's disease'; 346.
'Case of Addison's disease. Bronzing of the skin for 8 years.
Atrophy of the supra-renal capsules.'

Gulliver, G. (1840). *Dublin Med. Press.* **3**, 10–11. 'Notice of Mr. Gulliver's observations on the thymus and mesenteric glands; on the chyle; and on the supra-renal glands.'

Hédon, E. (1893). *Arch. physiol. norm. et path.* Vth series, **5**, 154–63. 'Sur la consommation du sucre chez le chien apres l'extirpation du pancréas.'

Hopkins, F. G. (1938). *Ergbn. Vit. Hormonforsch*, **1**, v–vi. Foreword (to the volume).

Houssay, B. A. (1952). *Diabetes*, **1**, 112–16. 'The discovery of pancreatic diabetes—the role of Oscar Minkowski.'

Lower, R. (1670). *Tractatus de corde.* Quoted by Rolleston, H. D. (1935), pp. 12 and 14.

von Mering, J. & Minkowski, O. (1889–90). *Arch. exp. Path. Pharm.* **26**, 371–87. 'Diabetes mellitus nach Pancreasexstirpation.'

Meyer, J. de (1909). *Arch. Fisiol.* **7**, 96–9. 'Action de la sécrétion interne du pancréas sur différents organes et en particulier sur la sécrétion rénale.'

Myers, V. C. & Bailey, C. V. (1916). *J. Biol. Chem.* **24**, 147–61. 'The Lewis and Benedict method for the estimation of blood sugar, with some observations obtained in disease.'

Oliver, G. & Schäfer, E. A. (1894). *J. Physiol.* **16**, i–iv (Proc.) 'On the physiological action of extract of the supra-renal capsules.'

Opie, E. L. (1900–1). *J. Exp. Med.* (N.Y.) **5**, 527–40. 'The relation of diabetes mellitus to lesions of the pancreas. Hyaline degeneration of the islands of Langerhans.'

Rolleston, H. D. (1936). *The endocrine organs in health and disease.* Oxford University Press, London.

Sanger, F. (1960). *Brit. med. Bull.* **16**, 183–8. 'Chemistry of insulin.'

Shaffer, P. A. & Hartman, A. F. (1921). *J. biol. Chem.* **45**, 365–90. 'Methods for the determination of reducing sugars in blood, urine and other solutions.'

Starling, E. H. (1905). *Lancet*, II, (*a*) 339–41; (*b*) 423–5; (*c*) 501–3; (*d*) 579–83. The Croonian Lectures on 'The chemical correlation of the functions of the body' given before the Royal College of Physicians of London on 20, 22, 27 and 29 June 1905. Lecture I (*a*) 'The chemical control of the functions of the body.' Lecture II (*b*) 'The chemical reflexes of the alimentary tract.' Lecture III (*c*) 'The chemical reflexes of the alimentary tract.' Lecture IV (*d*) 'The chemical correlations involving growth of organs.'

Swingle, W. W. & Pfiffner, J. J. (1930). *Science*, **71**, 321–2. 'An aqueous extract of the suprarenal cortex which maintains the life of bilaterally adrenalectomized cats.'

Todd, R. B. (1836–59). Editor of *Cyclopaedia of anatomy and physiology*. Vols. i–iii. Sherwood Gilbert and Piper, London. Vol. iv and Vol. v (Supplementary vol.), Longman, Brown, Green, Longmans and Roberts, London. Vol. i, 1836; Vol. ii, 1839; Vol. iii, 1847; Vol. iv, 1852; Vol. v, 1859.

Young, F. G. (1957). *Brit. med. J.* **1**, 1431–7. 'Claude Bernard and the discovery of glycogen. A century of retrospect.'

Young, F. G. (1959). In 'Significant trends in medical research'. Ciba Foundation Symposium on Medical Research (eds. G. E. W. Wolstenholme & M. O'Connor) London, Churchill. pp. 135–57. 'The nature and mechanism of action of hormones.'

Young, F. G. (1964). *Advancement of Science*, **21**, 369–78. 'Biochemistry of the endocrine system.'

Zuelzer, G. L. (1908). *Zeit. exp. Path. Therap.* **5**, 307–18. 'Ueber Versuche einer specifischen Fermenttherapie des Diabetes.'

6

THE DISCOVERY OF VITAMINS

by Leslie J. Harris

PRELIMINARY NOTIONS

During the course of the nineteenth century chemists gradually devised methods for analysing foodstuffs, and ascertained that they were composed essentially of three main classes of organic substances, proteins, fats and carbohydrates, together with various mineral salts and water. The components just enumerated were shown to account, quantitatively, for approximately one hundred per cent. of the chemical analysis; and it was natural to assume therefore that they were all that mattered. Thus, at the beginning of the twentieth century it was the accepted practice to define the nutritional value of any foodstuff in terms of these substances.

But, then, during the first quarter of the present century—following upon researches by Lunin (1881) in Switzerland, Eijkman (1890) in Java, and notably Hopkins (1912) in England, to mention only some of the more outstanding names—it became apparent that there were other things that mattered nutritionally in foods as well as these well-recognized main components. The newly detected ingredients, present in foods in extremely minute amounts, but nevertheless essential for the maintenance of health, came to be known as vitamins. Actually, the term 'vitamine' (spelled originally with a terminal -e) was coined by Funk (1912), at that time working at the Lister Institute in London. The word was intended by him to imply first that they were 'vital' for the survival of the living organism, and secondly that some of them at least were basic in chemical character, or 'amines'.

At that time, in 1912, no vitamin had yet been isolated, and their chemical nature was quite unknown. Little by little, one vitamin after another became recognized, each responsible for preventing some different disorder in man or animal. Later, one by one, they were to be separated in a pure state; their chemical nature established; their synthesis accomplished in the laboratory; their biochemical mode of action ascertained; methods devised for estimating the amounts present in various foodstuffs (raw, cooked or processed); and the actual requirements determined for human beings and other animals. Today more than a score of vitamins are known (many of them existing in more than one different form or modification); and at least a dozen of them have been shown to be needed by man, or to be of undoubted clinical significance in some set of circumstances or another.

Before we turn to a more detailed exposition of how history gradually unfolded itself, it may be well to attempt a definition (Harris, 1951) of what we understand by the term vitamin: 'Vitamins are substances that (*a*) are distributed in foodstuffs in relatively minute quantities, that (*b*) are distinct from the main components of food (i.e. proteins, fats, carbohydrates, mineral salts and water), that (*c*) are needed for the normal nutrition of the animal organism, and (*d*) the absence of any one of which causes a corresponding specific deficiency disease.'

VITAMIN HISTORY

From what has already been hinted it will be understood that recognition of the existence of vitamins as such has grown gradually. Actually, the first step, in a long series of developments leading eventually to the 'vitamin(e) hypothesis', may be discerned in some early observations on deficiency diseases in man, especially on scurvy and beri-beri.*

* For a more detailed history of the discovery of vitamins, see Harris (1935, 1938). The present text is based in part on the opening lecture in a course on Vitamins given to the Part II Tripos Class in Biochemistry at Cambridge, which formed the subject-matter of an earlier book (Harris, 1951), approval to quote freely from which has kindly been granted by the publishers, Messrs. J. and A. Churchill Ltd., to whom sincere thanks are hereby accorded.

Leslie J. Harris

Table 1. Chronological chart: the early history of vitamins

Part I. Deficiency diseases

In Man:

1601	Scurvy (Lancaster)	⎫
1882	Beri-beri (Takaki)	⎬ Prevented empirically by dietary additions.
c. 1900	Rickets	⎭

In experimental animals:

1890–7	Eijkman	Experimental beri-beri discovered. First work on anti-beri-beri factor.
1901	Grijns	Beri-beri simply a deficiency disease.
1907–12	Holst and Frölich	Experimental scurvy. Scurvy similarly a deficiency disease. Work on anti-scurvy factor.

Theory:

1840	Budd	Anti-scorbutic factor postulated.
1906	Hopkins	Scurvy and rickets—'Minimal qualitative factors'.
1912	Funk	The 'vitamine' theory. Anti-beri-beri, anti-scurvy, anti-rickets and anti-pellagra 'vitamines' postulated.

Part II. Normal diets

1881	Lunin	Purified basal diets inadequate.
⎧ 1909	Stepp	Extracted bread and milk inadequate.
⎨ 1905	Pekelharing	*Small* supplement of milk suffices.
1912	Hopkins	Convincing quantitative evidence for these accessory factors.
1915	McCollum and Davis	Two such factors at least.

Empirical cures of scurvy, beri-beri, and rickets

We have to go back to the sixteenth century for what appear to be the first records of the cure of scurvy—that old scourge of mariners—by such agents as: a decoction of spruce needles (Jacques Cartier, 1535), or oranges and lemons (Sir Richard Hawkins, 1593). It was early in the seventeenth century (1601) that Sir James Lancaster introduced the regular use of oranges and lemons into the ships of the East India Company as a preventive against the disease. Many others, during the seventeenth and eighteenth centuries repeatedly confirmed the fact that fresh fruits and vegetables were effective in curing or preventing scurvy (e.g. Woodall, 1639; Kramer, 1739; Lind,

158

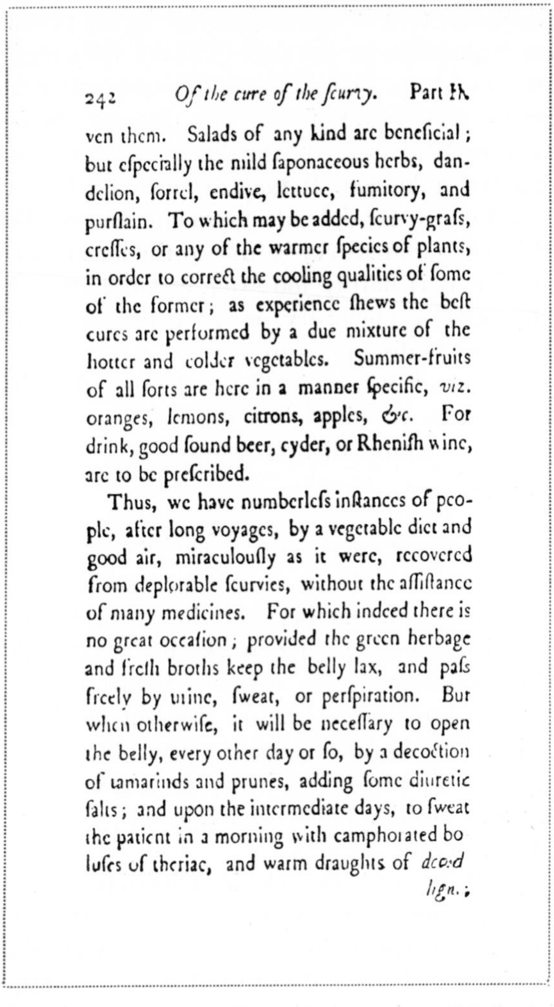

242   *Of the cure of the scurvy.*   Part II.

ven them.   Salads of any kind are beneficial ; but especially the mild saponaceous herbs, dandelion, sorrel, endive, lettuce, fumitory, and purslain. To which may be added, scurvy-grass, cresses, or any of the warmer species of plants, in order to correct the cooling qualities of some of the former; as experience shews the best cures are performed by a due mixture of the hotter and colder vegetables. Summer-fruits of all sorts are here in a manner specific, *viz.* oranges, lemons, citrons, apples, &c. For drink, good sound beer, cyder, or Rhenish wine, are to be prescribed.

Thus, we have numberless instances of people, after long voyages, by a vegetable diet and good air, miraculously as it were, recovered from deplorable scurvies, without the assistance of many medicines. For which indeed there is no great occasion ; provided the green herbage and fresh broths keep the belly lax, and pass freely by urine, sweat, or perspiration. But when otherwise, it will be necessary to open the belly, every other day or so, by a decoction of tamarinds and prunes, adding some diuretic salts; and upon the intermediate days, to sweat the patient in a morning with camphorated boluses of theriac, and warm draughts of *decod lign.;*

Fig. 1. Extract from Lind's *A treatise on the scurvy*, 1753. (Obtained through the services of the British Museum Reading Room)

1757; Captain James Cook, 1772). In 1804 the daily consumption of lemon juice was made compulsory in the British Navy.

In 1882 Takaki found that he could stamp out beri-beri in the Japanese Navy by certain changes in the diet. He considered, however, that it was an increase in the protein intake which was responsible for the cure (Takaki, 1885 etc.).

Thus we may say that although dietary cures had been discovered for the control of these two diseases—beri-beri and scurvy—yet the true nature of the dietary error responsible for them remained unknown. It may nevertheless be recalled in parenthesis that a far-sighted physician named Budd had predicted in 1840 that scurvy is 'due to the lack of an essential element which it is hardly too sanguine to state will be discovered by organic chemistry or the experiments of physiologists in a not too distant future'.

Towards the end of the nineteenth century the view began to be expressed by paediatricians that 'rickets is produced as certainly by a rachitic diet as scurvy by a scorbutic diet' (Cheadle, 1899).

Discovery of experimental avitaminoses

In 1890 Eijkman, in the Dutch East Indies, made the important discovery of experimental beri-beri in fowls. From 1890 to 1897 he carried out the earliest work on the extraction of the antineuritic substance (now called vitamin B_1), which he found to be present in the bran of rice but not in 'polished' (milled) rice. But the first to state clearly that beri-beri was due solely and simply to a dietary deficiency and was not caused by any positive agent or toxin, was Eijkman's collaborator, Grijns (1901).

In 1907, Holst and Frölich in Christiania [now Oslo] discovered experimental scurvy in guinea-pigs. With Eijkman's pioneer work on beri-beri in mind they rightly considered this to be likewise a deficiency disease and set out to examine the properties of the anti-scorbutic substance (now vitamin C).

The concept of vitamins

By 1906 Hopkins could refer to scurvy and rickets as 'diseases in which for long years we have had knowledge of a dietetic factor'. He realized, moreover, that the errors in the diet 'although still obscure' were 'certainly of the kind which comprises the minimal qualitative factors'. In 1912 Funk,

Plate 1. Captain James Lind, author of *A treatise on the scurvy*, 1753 (from a paper by Sir H. Rolleston in *J. Royal Navy Med. Services*, 1915)

Plate 2. The Prize Medal awarded by the Royal Society in 1776 to Captain James Cook, who, in his book *Voyage towards the South Pole and round the World*, describes the dietetic measures taken to prevent the occurrence of scurvy in his sailors. (Obtained through the services of the British Museum Reading Room)

Plate 3. Admiral K. Takaki, who eradicated beri-beri from the Japanese navy, 1882. (From the Japan Gazette *Peerage of Japan* obtained by courtesy of the Trustees of the British Museum)

Plate 4. Professor C. Eijkman. He was the first to produce a vitamin-deficiency disease in an experimental animal (1890–7), and in 1929 was awarded a Nobel Prize jointly with Hopkins (plate 6). (Obtained by the services of the Archivist of the University of Utrecht)

Plate 5. Dr Casimir Funk, who propounded the 'vitamine' hypothesis in 1912

Plate 6. Sir Frederick Gowland Hopkins O.M., F.R.S. (after the Royal Society Portrait by Meredith Frampton, R.A.) He was awarded the Nobel Prize in Medicine for 1929, jointly with Eijkman (plate 4), for his experiments on the 'accessory factors' in food. (Reproduced by kind permission of the artist)

then working on the anti-beri-beri factor, propounded his 'vitamine' theory—i.e. he postulated the existence of separate anti-beri-beri, anti-scurvy, anti-rickets, and anti-pellagra 'vitamines'.

Experiments on 'synthetic diets'

In the meantime, experiments had been in progress attacking the problem from an entirely different angle—investigating not deficiency diseases as such, but determining what constituted a physiologically complete diet. Lunin, a pupil of the Swiss biochemist Bunge, first showed in 1881 that animals failed to thrive when kept on an artificial regimen comprising the then known constituents of food, i.e. re-purified fat, protein, carbohydrate, mineral salts, and water. He concluded that 'a natural food such as milk must therefore contain besides these known principal ingredients small quantities of other and unknown substances essential to life'. Similar conclusions were reached by various other workers, of whom the most notable were Socin (1891) and Stepp (1909); mention must also be made of Coppola (1890), Hall (1896), Häusermann (1897), Henriques and Hansen (1905), Falta and Noeggerath (1905), and Jacob (1906). The eminent Dutch physiologist Pekelharing published, in 1905, the statement—generally overlooked at the time—that the unknown substances must be effective in very minute amounts, for he had found that quite small supplements of natural foods added to the artificial synthetic ration were sufficient to afford protection. An independent and more detailed study by Hopkins (1912) had as its principal conclusions: (i) that an insignificantly small addition of milk would suffice to render the purified diet adequate; and (ii) that the animals ceased to grow while still eating sufficient *in quantity* of the purified diet to support good growth (see fig. 2).

THE MODERN PERIOD

The year 1912 may be said to mark the beginning of modern intensive work on vitamins. Hopkins' celebrated paper and Funk's review, published a few months earlier, for the first time attracted world-wide attention to 'the vitamin question'.

Leslie J. Harris

For a few years longer, nevertheless, the existence of vitamins was still disputed.

In 1929 the importance of the pioneer experiments of Eijkman and of Hopkins was recognized by the award to them jointly of the Nobel Prize for Medicine.

Differentiation of fat-soluble and water-soluble factors

'Fat-soluble A' and 'water-soluble B'. The first real indication of the multiplicity of vitamins came in 1915 when McCollum and Davis in America showed that, with the rat as an experimental animal, at least two accessory factors were needed for growth. One was present in fatty and the other in non-fatty foods. These were named respectively 'fat-soluble A' and 'water-soluble B'.

The beginnings of the vitamin alphabet

Vitamin B. Before long it became recognized that what was called 'water-soluble B' had the properties of the antineuritic (anti-beri-beri) 'vitamine'. To avoid confusion the two systems of nomenclature were combined. At the suggestion of J. C. Drummond (1920) 'water-soluble B' was re-named 'vitamin B', the terminal 'e' of the word vitamine being omitted to avoid the possibly unwarranted implication that it necessarily had the chemical nature of an amine.

Vitamin A. Similarly 'fat-soluble A' was re-named 'vitamin A'. The most noticeable effects of its deficiency were found to include xerophthalmia, failure of growth and increased liability to infection, especially of the respiratory system.

Vitamin C. It was apparent that the anti-scurvy factor had entirely different chemical properties from either 'vitamin B' or 'vitamin A'. It was accordingly given the next letter in the alphabet and became 'vitamin C'.

The growth of the alphabet

Vitamin D. A little later it was proved conclusively that experimental rickets in dogs was due to the absence of a fat-soluble vitamin (Mellanby, 1918). At first it was thought to be identical with vitamin A, already known to be present in the curative agent, cod-liver oil. Later, however, differences in distribution and chemical properties were established (McCollum, Simmonds, Becker & Shipley, 1922) and it was accordingly named vitamin D.

Vitamin E. A new factor, vitamin E, needed to ensure normal reproduction in the rat, was described by Evans and Bishop in California in 1922. Independent evidence of this new anti-sterility vitamin was obtained almost simultaneously by Sure and by Mattill, also in America.

'*Vitamin B*' as a complex

Vitamin B_1 and vitamin B_2. In 1926 Goldberger and his colleagues proved, as Funk had predicted, that pellagra was associated with the lack of a vitamin. He showed that it had a distribution somewhat similar to that of the anti-beri-beri factor but was more stable to heat. He called it the P.-P. (pellagra-preventing) factor. An official committee in England (Accessory Food Factors Committee, 1927), however, revised the nomenclature. The anti-beri-beri vitamin was re-named vitamin B_1, and the 'heat-stable component'—then assumed to be a single substance—was called 'B_2'. 'Vitamin B_2' was defined as the heat-stable, water-soluble factor, present in yeast extracts, needed to prevent pellagra in man, and the apparently similar diseases in dogs (*canine black-tongue*) and in rats (*pellagra-like dermatitis*), and needed also for promoting the growth of rats. 'Vitamin B_2' itself, as so defined, however, soon proved to be complex.

'*Vitamin B₂*' as a complex

Riboflavin. The first component of the B_2 complex to be characterized was the substance riboflavin, previously known as a naturally occurring yellow pigment in milk, which in 1933 was shown to possess growth-promoting activity for rats (Kuhn, György & Wagner-Jauregg, 1933). Riboflavin is, therefore, sometimes still referred to simply as 'vitamin B_2'.

It was later found that a deficiency of riboflavin is the cause of 'cheilosis' (lesions on the lips) in man.

Vitamin B_6 (adermin, pyridoxin). Soon afterwards, in 1934, a second component of the vitamin B_2 complex, at first called vitamin B_6 (György),* but later re-named adermin, and now generally known as pyridoxin, was identified, also by experiments on rats. Its absence from their diet caused severe skin lesions.

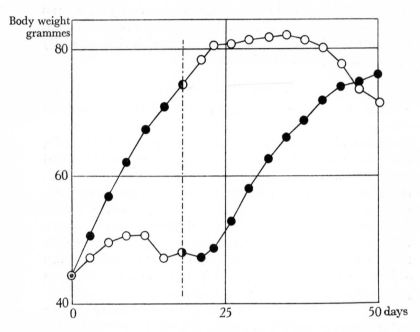

Fig. 2. Weight curves of growing rats with and without dietary supplement (Hopkins, 1912). ○ Artificial diet alone. ● Artificial diet plus small daily addition of milk

Nicotinamide. In 1935, Birch, György and Harris showed that the pellagra-preventing factor proper was a third component, distinct from both riboflavin or vitamin B_6. In 1937, this 'P.-P.' vitamin was identified (Elvehjem *et al.*, 1937) with nicotinic acid, or nicotinamide (now generally known in America as niacin, and niacin amide, respectively).

* The name B_6 was chosen because the intervening symbols, B_3 to B_5, had previously been used in other connexions—although they have since fallen into desuetude.

The discovery of vitamins

Other vitamins

Year by year additional vitamins were coming to light, as listed below. Three at least of these have proved of definite importance in human nutrition, viz., vitamin K, folic acid, and vitamin B_{12}. Several others are considered, at any rate in some particular set of circumstances, to be needed by man. A number of the remainder have so far been studied, mostly on experimental animals, and their exact significance for the human being thus remains undecided. It should be mentioned too that several of these vitamins are also needed as growth factors by micro-organisms.

The following list is approximately in chronological order, but is not intended to be comprehensive:

(*a*) Vitamin F. Nutritionally essential unsaturated fatty acids (linoleic, linolenic and arachidonic acids) needed by rats (Burr & Burr, 1929; Evans, Lepkovsky & Murphy, 1934).

(*b*) Pantothenic acid. Prevents a specific dermatitis in chicks, formerly called, somewhat misleadingly, 'chick pellagra' (Norris & Ringrose, 1930), and is identical with a so-called filtrate factor for rats. Its deficiency in man may, apparently, sometimes be the cause of glossitis, and perhaps of an endemic disease ('burning-feet' syndrome) seen in Southern India.

(*c*) Vitamin H or biotin. First known as the 'H' or 'Haut' (= skin) factor for rats (György, 1931), biotin, as it is now called, has been shown to be needed by human beings, although it has not yet found any regular use in clinical medicine. In chicks, likewise, a deficiency of biotin has been found to cause dermatitis.

(*d*) Choline. Concerned in fat metabolism in rats (Best *et al.*, 1932, 1935), and a biological transmethylating agent. A deficiency in rats causes abnormalities in liver, kidneys and elsewhere.

(*e*) Vitamin K. This was first recognized by Dam in 1935 as concerned in preventing haemorrhages in chicks, and was later shown to have clinical applications in the cure of certain haemorrhagic disorders due to a condition of hypoprothrombinaemia. In other words, vitamin K is needed for the elaboration in the liver of the prothrombin, concerned in the clotting of the blood.

(*f*) Vitamin P. Supposed 'permeability' (= capillary fragility) factor for guinea-pigs (Rusznyak & Szent-Györgyi, 1936): possible human relations still in discussion.

(*g*) Folic Acid (pteroyl-glutamic acid). This factor, whose recognition dates from about 1941, is of interest in connexion with certain types of anaemia.

(*h*) Inositol. Needed for the growth of various micro-organisms (Eastcott, 1928), and said to prevent loss of hair in rats and mice (Woolley, 1940*a*, *b*; Pavcek & Baum, 1941; Nielson & Black, 1944).

(*i*) *p*-Aminobenzoic acid. Growth factor for certain micro-organisms; its special interest is that it is the specific bacterial nutrient that is antagonized by the sulphonamide type of drug (Woods, 1940). Considered to be needed to maintain a normal pigmentation of the hair—i.e. to prevent it greying—in some species (Ansbacher, 1941).

(*j*) Strepogenin. Growth factor for certain haemolytic streptococci; found also to promote growth in mice and guinea-pigs (Woolley, 1941) and perhaps in rats.

(*k*) Animal-protein factors. Associated with proteins of animal origin, and stimulating the growth of animals restricted to certain experimental diets (Cary *et al.*, 1946; Hartman, 1946); see following entry:

(*l*) Vitamin B_{12}. One of the 'animal-protein factors', first investigated as a growth factor for a particular micro-organism (Shorb, 1947), and almost immediately afterwards identified as the anti-pernicious-anaemia factor. A point of special interest is that its molecule contains the mineral element, cobalt.

ADVANCES IN VITAMIN RESEARCH

Chemical constitution and synthesis

During the decade 1928 to 1938, as a result of intensive effort by hundreds of investigators, the better known vitamins (including A, B_1, C, D and E) were isolated in a state of purity, one by one, and their chemical constitutions were then worked out. Within a few years most of these (vitamins A, B_1, C and E) had been synthesized in the laboratory, and the 'artificial' product proved to be identical in properties and physiological effect with the 'natural' one (table 2).

Table 2. Progress in vitamin chemistry: the four 'classical deficiency diseases'

Deficiency disease	Discovery of protective foodstuff	Recognition of a specific protective factor	Vitamin isolated (or identified with substance previously known)*	Vitamin synthesized (or synthesized before known to be a vitamin)†
Scurvy	Fruits and vegetables (c. 1600)	1907	(1932)*	1933
Beri-beri	Whole rice, barley etc. (1882)	1901	1927	1936
Rickets	'Good fats' (c. 1900)	1919–22	1932	—‡
Pellagra	'Good protein foods' (1916)	1926	(1937)*	(1867)†

‡ Vitamin D_2 was prepared 'artificially', by irradiation of the provitamin, ergosterol, in 1927. The 'complete synthesis' had still to be accomplished.

Later, other vitamins were in turn to be similarly characterized and synthesized—or, in some cases (e.g. nicotinic acid), they were proved to be identical with substances which had previously been known to chemists but had not yet been recognized to possess vitamin activity.

Chemical names and chemical relations

Vitamin A ($C_{20}H_{29}OH$) was shown to be chemically related to the naturally occurring hydrocarbon β-carotene ($C_{40}H_{56}$), which shares its biological activity. Vitamin B_1 ($C_{12}H_{17}N_4OSCl$. HCl) was given the name of aneurin, or, later, thiamine. Vitamin C, ascorbic acid, has the formula $C_6H_8O_6$. There are several forms of vitamin D, the most important being calciferol or 'vitamin D_2' ($C_{28}H_{43}OH$) and 'vitamin D_3'

$(C_{27}H_{43}OH)$. Vitamin E, α-tocopherol, has the formula $C_{29}H_{50}O_2$; there are also β-, γ- and δ-tocopherols, with somewhat lower biological potency (at any rate for rats). Chemical names and synonyms for other vitamins have likewise been adopted.

Further progress

Side by side with these investigations on the organic chemistry of the vitamins, methods were being elaborated by which vitamins could be accurately assayed in biological materials or in footstuffs, and the losses incurred on cooking or processing ascertained; their chemical and physical properties were being examined in close detail; tests were devised to assess the status of a human subject in some particular vitamin; the daily requirements were being estimated; some new and unexpected clinical uses were found; and the mode of action of the vitamins in the body was being gradually elucidated.

A notable advance in this last-mentioned direction was the discovery that various vitamins of the B group owe their biological activity to the fact that they can function as co-enzymes in the living cell. The pioneer observation of this kind was the finding of Lohmann and Schuster in 1937 that the pyrophosphate ester of vitamin B_1 is the co-enzyme for decarboxylation of pyruvic acid, a substance which is an important metabolite in the breakdown of carbohydrates.

Recapitulation

The early stages of vitamin history are summarized in tables 1 and 2. For a fuller catalogue of various other vitamins now known, reference may be made again to the foregoing text (pp. 162 *et seq.*).

The discovery of vitamins

REFERENCES

Accessory Food Factors Committee (1927). Minutes of Meeting, 14 November.

Ansbacher, S. (1941). *Science*, **93**, 164.

Best, C. H., Hershey, J. M. & Huntsman, M. E. (1932). *Amer. J. Physiol.* **101**, 7; *J. Physiol.* **75**, 56.

Best, C. H., Huntsman, M. E., McHenry, E. W. & Ridout, J. H. (1935). *J. Physiol.* **84**, 38P.

Birch, T. W., György, P. & Harris, L. J. (1935). *Biochem. J.* **29**, 2830.

Budd, G. (1840). *Tweedie's system of practical medicine.* Philadelphia, p. 99.

Burr, G. O. & Burr, M. M. (1929). *J. biol. Chem.* **82**, 345.

Cartier, J. (1535). Quoted by Major, R. H. (1932). *The doctor explains.* Chapman and Hall: London.

Cary, C. A., Hartman, A. M., Dryden, L. P. & Likely, G. D. (1946). *Fed. Proc.* **5**, 128.

Cheadle, W. B. (1889 etc.). *Artificial feeding of infants.* London, 1st edition, 1889; subsequent editions, 1892, 1894, 1896, 1902, 1906.

Cook, J. (1772–5). *Voyage towards the South Pole and round the World.*

Coppola, F. (1890). *R. C. Accad. Lincei*, **6** (i), 362.

Dam, H. (1935). *Nature, Lond.* **135**, 652; *Biochem. J.* **29**, 1273.

Drummond, J. C. (1920). *Biochem. J.* **14**, 660.

Eastcott, E. V. (1928). *J. phys. Chem.* **32**, 1094.

Eijkman, C. (1890). *Geneesk Tijdschr. Nederland-Indie*, **30**, 295.

Elvehjem, C. A., Madden, R. J., Strong, F. M. & Woolley, D. W. (1937). *J. Amer. chem. Soc.* **59**, 1767.

Evans, H. M. & Bishop, K. S. (1922). *Science*, **56**, 650.

Evans, H. M., Lepkovsky, S. & Murphy, E. A. (1934). *J. biol. Chem.* **106**, 431, 441, 445.

Falta, W. & Noeggerath, C. T. (1905). *Beitr. chem. Physiol. Path.* **7**, 320.

Funk, C. (1912). *J. State Med.* **20**, 341.

Goldberger, J., Wheeler, G. A., Lillie, R. D. & Rogers, L. M. (1926). *Publ. Hlth. Rep., Wash.* **41**, 297.

Grijns, G. (1901). *Geneesk. Tijdschr. Nederland-Indie*, **41**, 3.

György, P. (1931). *Z. ärztl. Fortbildung.* **28**, 377, 417.

György, P. (1934). *Nature, Lond.* **133**, 498.

Hall, W. S. (1896). *Arch. Anat. Physiol., Lpz.*, Physiol. Abth., 142.

Harris, L. J. (1935). *Vitamins in theory and practice.* Cambridge, 1st ed.

Harris, L. J. (1938). *Vitamins and vitamin deficiencies*, Vol. 1, *introduction and historical; vitamin B_1 and beri-beri.* London.

Harris, L. J. (1951). *Vitamins: a digest of current knowledge.* London.

Hartman, A. M. (1946). *Fed. Proc.* **5**, 137.

Häusermann, E. (1897). *Hoppe-Seyl. Z.* **23**, 555.

Hawkins, Sir Richard (1593). *Observations on his voyage to the South Sea.*

Henriques, V. & Hansen, C. (1905). *Hoppe-Seyl. Z.* **43**, 417.

Holst, A. & Frölich, T. (1907). *J. Hyg. Camb.* **7**, 634.

Hopkins, F. G. (1906). *Analyst,* **31**, 395.

Hopkins, F. G. (1912). *J. Physiol.* **44**, 425.

Jacob, L. (1906). *Z. Biol.* **48**, 19.

Kramer. J. G. H. (1739). Medicina Castrensis, Wien.

Kuhn, R., György, P. & Wagner-Jauregg, T. (1933). *Ber. dtsch. chem. Ges.* **66**B, 317, 576.

Lancaster, J. (1601). Quoted by Nixon, J. A. (1937). *Proc. R. Soc. Med.* **31**, 193.

Lind, J. (1757). *A treatise on the scurvy.* London, 2nd ed.

Lohmann, K. & Schuster, P. (1937). *Biochem. Z.* **294**, 188.

Lunin, N. (1881). *Hoppe-Seyl. Z.* **5**, 31.

McCollum, E. V. & Davis, M. (1915). *J. biol. Chem.* **23**, 181, 231.

McCollum, E. V., Simmonds, N., Becker, J. E. & Shipley, P. G. (1922). *J. biol. Chem.* **53**, 293.

Mattill, H. A. & Conklin, R. E. (1920). *J. biol. Chem.* **44**, 137.

Mellanby, E. (1918). *J. Physiol.* **52**, proc. pp. xi, liii.

Nielson, E. & Black, A. (1944). *Proc. Soc. exp. Biol., N.Y.* **55**, 14.

Norris, L. C. & Ringrose, A. T. (1930). *Science,* **71**, 643.

Pavcek, P. L. & Baum, H. M. (1941). *Science,* **93**, 502.

Pekelharing, C. A. (1905). *Ned. Tijdschr. Geneesk.* **70**, 111.

Rusznyak, S. & Szent-Gyorgyi, A. (1936). *Nature, Lond.* **138**, 27.

Shorb, M. S. (1948). *Science,* **107**, 397.

Socin, C. A. (1891). *Hoppe-Seyl. Z.* **15**, 93.

Stepp. W. (1909). *Biochem. Z.* **22**, 452.

Sure, B. (1923–4). *J. biol. Chem.* **58**, 681, 693.

Takaki, K. (1885). *Sei-I-Kwai med. J.* **4**, supplement, 29.

Woodall, J. (1639). *The surgeon's mate, or military and domestique surgery.* London.

Woods, D. D. (1940). *Brit. J. exp. Path.* **21**, 74.

Woolley, D. W. (1940 a). *J. biol. Chem.* **136**, 113.

Woolley, D. W. (1940 b). *Science,* **92**, 384.

Woolley, D. W. (1941). *J. exp. Med.* **73**, 487.

7

THE HISTORICAL FOUNDATIONS
OF MODERN BIOCHEMISTRY

by Mikuláš Teich

Biochemistry matured into its modern stage at the close of the nineteenth century. Various influences of a philosophical, scientific and social nature contributed towards its making and within this short chapter it will, of course, be impossible to deal with the subject exhaustively. Only an outline of the consecutive stages of development can be attempted here.

It seems to me that the rise of modern biochemistry involved a gradual separation of the science of the chemistry of life from the general body of chemistry to form organic chemistry (from about 1800 to 1840). This was followed by the linking of organic chemistry with physiology (about 1840–80). Finally, modern biochemistry as we know it today separated from physiology.

VITAL FORCE

The Lavoisierian revolution in chemistry which was the outcome of a complex historical process transformed chemistry into science. It was now possible to begin to understand in a rational manner, chemical changes which confronted man in nature and industry. One of the prerequisites for an understanding of chemical changes in living bodies was greater knowledge about the nature and composition of substances of biological origin. Leading pioneers in this field were the French chemists Fourcroy and Vauquelin who were interested in gaining knowledge of these substances mainly from the medical point of view.

Investigations of these and other workers showed that, essentially, there was no difference chemically between bodies of mineral and biological origin. This was already the view of

Lavoisier who applied his oxidation theory equally to inorganic and organic compounds. Inorganic bodies contained elements which were to be found by analysis in organic bodies. Fourcroy and others became protagonists of the view that inorganic compounds were either binary, that is they contained two elements, or were derivatives of binary substances; organic compounds were thought to be of ternary or quarternary nature.

Carbonic acid and ammonia were binary compounds, the first being composed of carbon and oxygen, the second of nitrogen and hydrogen. These two binary compounds were able in turn to form a new inorganic compound, namely ammonium carbonate. Sugar, starch and fats were considered to be ternary compounds composed of carbon, oxygen and hydrogen, and gluten and albumen to be quarternary substances containing nitrogen in addition. Since it was generally accepted that there was no chemical difference between bodies of inorganic and organic origin, the question naturally arose: what was the basis for the phenomenon that the same elements in certain cases were able to produce inorganic compounds and in others organic compounds? The explanation was found in the existence of a vital force.

It appears that the term was coined in 1774 by Friedrich Casimir Medicus, but the first systematic treatment of it came twenty years later, clearly, as a result of the changes which chemistry was undergoing at that time. In July 1795 the first issue of the *Archiv für die Physiologie* appeared; it was published by Johann Christian Reil who was then at Halle University. The new journal opened with an article by Reil 'Von der Lebenskraft' (On the Vital Force). Reil's exposition is a very fine example of how to blend experimental knowledge with theoretical thought and there is little evidence of that flight of fantasy which characterizes some of the extreme representatives of German Naturphilosophie. Reil emphasized that the phenomena of animal life cannot be deduced from ideas which are not confirmable by experience (Reil, pp. 10–11). His main theoretical thesis was that the phenomena which are displayed by living bodies are materially conditioned. He says:

The foundations of modern biochemistry

That we do not find in inanimate nature phenomena belonging to those of the animate world depends on the particular kind of organic matter that cannot be found in inanimate nature. Can we, however, deny to all others special qualities which a certain kind of matter lacks? Is it necessary to deduce the magnetic quality of iron from something other than matter because we cannot observe magnetic qualities in tin, in stones and in wood? (Reil, p. 14).

Reil was convinced that all the manifold phenomena in nature which were encountered in non-living and living bodies could, in the last analysis, be reduced to problems of mixture and the form of matter. Again and again he returns to this point and quite naturally emphasizes the role of chemistry. 'If matter and its nature is the basis of all phenomena of living bodies', he writes, 'then the chemical analysis of organic bodies will grow ever more important to theoretical and practical medicine. However, our science of mixtures of organic bodies is still far from perfect' (Reil, p. 25). It may be noticed that so far Reil had not introduced the concept of the vital force into his deliberations. He simply did not need it. At one point he says: 'Force is a subjective concept, the form in which we think of the relations between cause and effect. If it were possible for us to think clearly of each body as it is, of the nature of all its constituents and their combination, and of their mixture and form, then we would not need the concept of force, which gives rise to quite a number of erroneous conclusions' (Reil, p. 46). On reading the section on force, one is struck by how unhappy its author was when he had to fall back on using this term. This is shown in the numerous explanatory footnotes which accompany his otherwise very clear text. In one of them he says:

I think we would give rise to the least possible misconception if, instead of force, we speak of the *property of matter*. In physiology we would then just consider general phenomena which organic matter has in common with inanimate nature. Then we would reflect on those of its special properties which belong to the whole animate kingdom of nature. Then we would pass over to the special phenomena associated with vegetable and animal matter (Reil, p. 47).

173

He defines vital force in the following way:

Lebenskraft indicates the relation of several individual phenomena to a special kind of matter which we encounter in living nature with plants and animals. The most general attribute of this particular kind of matter is a special kind of crystallization. Besides, we cannot offer a genetic definition of this force as long as chemistry does not inform us with greater precision about the basic elements of organic matter and their properties (Reil, p. 48).

In a footnote to this definition Reil is at pains to point out:

I have called *vital force* the force of matter which characterizes the animal and plant kingdom...Perhaps others will find the word *organic force* more appropriate. I did not choose it, however, because organization in ordinary usage signifies the formation of animate beings. But words are arbitrary signs of our concepts and it depends only on how precisely we define the concept which we associate with a certain word (Reil, p. 48).

From what has been shown it is quite evident that Reil's view on the vital force was anything but mysterious. He was a firm believer that various manifestations of life in due course would be explained in terms of chemistry and it would be possible to dispense with the essentially misty notion of vital force. This, however, did not happen.

It is only necessary to look into the various editions of the famous text-book of Berzelius to see that vital force was not dispensed with. In the first German edition which appeared in 1827 the section on organic chemistry starts as follows:

In living, as compared with inanimate nature elements appear to obey quite different laws. The products of their mutual reactions are quite different from those in the sphere of inorganic nature. The discovery of the cause of this difference of the behaviour of elements in living and non-living nature would furnish the key to the theory of organic chemistry. Meanwhile it is so concealed that, at least at present, we are without any hope of discovering it. In spite of this we must strive to get nearer to the knowledge of it, because one day we shall be lucky either to reach it completely or else attain a certain limit beyond which human power of investigation cannot go. A living body, considered as an object of chemical investigation, is a workshop in which many chemical processes are taking place with the result that they produce all the phenomena which we call life as a whole... (p. 135).

In this passage Berzelius appears not to rule out the possibility that chemistry will be able to say something more definite about the difference between the inanimate and animate. But it is necessary to remember that he was writing about elements and their behaviour in non-living and living bodies. The Swedish chemist differentiated between the behaviour of chemical elements in the living body and something which made some bodies come to life.

This *something* which we call vital force lies quite apart from inorganic elements and does not belong to one of their original properties such as gravity, impermeability, electrical polarity etc.; but what it is, how it comes into being, and how it passes away, we do not comprehend. It is possible to foresee that the earth, with all its inorganic constituents, but without the presence of life, would continue, under the same conditions, to exist without living creatures. A force, incomprehensible to us and foreign to inanimate nature, has once imbued the inorganic mass with this *something*, not in such a way as to appear as a product of an accident (p. 137).

Berzelius continues to say that everything has been arranged supremely wisely and with a certain end in view. Living creatures appear and pass away, these continuous successions and changes forming in effect the basis for their own existence. Berzelius did not change his views on this subject during his life, and, in fact, they were reproduced substantially unchanged in the fifth edition of his book, which appeared posthumously in 1856.

It would, of course, be misleading to state that all chemists dealing with problems of life shared the views of Berzelius. But it is true to say that even those who were prepared to discard vital force as a completely unscientific notion and to reduce phenomena of life to pure physics and chemistry were in the end unable to solve the problem. Vital force belongs to animas, entelechies, souls, archaei, which, as Needham puts it, dance processionally through the history of European thinking because a *deus* always had to be found for a *machina* (Needham, 1956).

In this connexion it is perhaps worth while to try and evaluate the historical significance of Wöhler's preparation of urea in 1828. Interest in this problem has been revived recently following McKie's conclusion that it is a legend that Wöhler's preparation of urea from ammonium cyanate rang the death-knell of vitalism in organic chemistry. His main thesis is that it is highly arguable whether one can talk about a synthesis, as is usually done, and that it would be better to talk about transformation of the ammonium cyanate into urea. McKie defines synthesis as 'the compounding of a substance from the elements that compose it' and points out that the cyanate as then prepared originated from organic matter. McKie considers the first synthesis to be the synthesis of acetic acid achieved by Kolbe in 1845. He further maintains that neither Wöhler himself, nor Berzelius, nor the leading French organic chemist Gerhardt, conceived Wöhler's preparation of urea as being a refutation of vitalism. McKie concludes that 'Vitalism in organic chemistry was rejected not by a sudden and dramatic synthesis—for science does not advance and Nature does not reveal herself in that way—but by steady accumulation of contradictory facts' (McKie, 1944). His article appeared in 1944 in *Nature* and was recently referred to approvingly by Bodenheimer in his book on the history of biology and Good-field in her discussion of the mechanist–vitalist controversy as illustrated by problems of respiration and animal heat (Boden-heimer, p. 67; Goodfield, p. 126).

Much of what McKie writes is true, yet his evaluation of Wöhler's *transformation* of ammonium cyanate into urea is not correct. In fact, the preparation of this substance revealed one of those *contradictory* facts which McKie thinks—and rightly so —helped to banish vitalism not only from organic chemistry but from science as a whole.

Which contradiction had been revealed by Wöhler's preparation of urea from ammonium cyanate? It is the one which Wöhler describes in his letter to Berzelius as follows: '... and I must tell you that I can prepare urea without requiring kidneys or an animal, either man or dog (Berzelius & Wöhler, 1901, p. 206). Wöhler was puzzled by this because he produced a substance artificially which was otherwise produced only by living creatures. He did not think that he

The foundations of modern biochemistry

drove vitalism out of organic chemistry because he was not particularly concerned with this philosophical question. It may well be that this, in fact, helped to increase his mental uncertainty as to how to interpret his experiment. But it would be entirely wrong to think that Wöhler or Berzelius did not realize the significance of the artificial production of urea. From Berzelius's answer to Wöhler it would appear that the Swedish chemist took Wöhler's communication rather lightly. But in fact he says: 'It is quite an important and nice discovery which Hr. Doktor effected and I was indescribably pleased to hear of it' (Berzelius & Wöhler, 1901, p. 208). Then he continues to discuss peculiarities of the reaction from a purely chemical point of view. He does the same in his textbook on animal chemistry which appeared in German in 1831 where he states very clearly: 'Wöhler made a remarkable discovery that urea can be produced *artificially* [our italics—m.t.]' (Berzelius, 1831, p. 356). Undoubtedly the artificial method of producing a substance which was ordinarily produced by a biological process created a puzzle which scientists like Berzelius had to face. He could not deny the fact and he did not want—or was not able—to face its implications. Therefore he suggested that urea was on the borderline between organic and inorganic substances, and that may have been the reason why it could be produced artificially. This view has been taken up by others who also were unable to resolve the puzzle. Thus J. Müller, the famous German physiologist, wrote in 1835 in the second edition of his well known *Handbuch der Physiologie des Menschen*: 'The way that elements combine in organic bodies is peculiar and conditioned by peculiar forces. Though chemistry can dissolve organic compounds it cannot create them' (Müller, 1835, p. 8). After rejecting claims of various authors to have produced artificial organic substances, Müller admits only the validity of Wöhler's discoveries, and then continues: 'However, urea is placed at the extreme border of organic substances and is more of an excretion than a component of the animal body. Perhaps urea is not at all a compound with the characteristic properties of organic products' (Müller, 1835, p. 9).

For a long time Wöhler's achievement remained an isolated example of its kind. Yet it is an important landmark in the history of biochemistry. The contradiction which it raised could be solved only by the development of both its chemical and biological aspects. In fact, the previously unified science of chemistry began to show definite signs of splitting into inorganic and organic chemistry: organic chemistry from occupying itself with chemical properties of substances of vegetable and animal origin became in the end the chemistry of carbon compounds. That involved a process of working out a theoretical basis which no longer had any need to fall back on the vital force concept. A different situation obtained with plant and animal physiology which studied the changes these substances were undergoing in living bodies. There the vital force controversy profoundly influenced explanations of the nature of these changes.

INORGANIC AND ORGANIC CHEMISTRY

Chemistry at the time of Lavoisier was conceived as a unified branch of natural science. From the purely chemical point of view, Berzelius, in developing the concepts of Lavoisier, saw the difference between inorganic and organic bodies as being that the former were oxides of simple radicals whereas the latter were oxides of compound radicals. With plants the compound radicals consisted in general of carbon and hydrogen, and with animals of carbon, hydrogen and nitrogen. The radicals were believed to retain their identity in chemical changes. In that sense a compound radical in organic chemistry took the place of an element in inorganic chemistry.

Berzelius was also a strong advocate of applying electrochemical concepts to both inorganic and organic compounds. Since elements were considered to be either electropositive or electronegative, the same could be thought about radicals. Metals, hydrogen and alcohol radicals were electropositive, whereas chlorine, benzoyl and other acid radicals were considered to be electronegative. By 1838 it became increasingly obvious that it was impossible to cover the facts of organic

178

chemistry on the basis of an electrochemical theory of radicals. One of the difficulties was that chlorine, generally accepted to be an electronegative element, was able to replace the electropositive hydrogen in some organic compounds. This fact became the basis for the development of the substitution theory initiated by the French chemist Dumas.

From the study of the correspondence between Berzelius and Liebig it is possible to get an intimate view of some of the problems which gave rise to the separation of organic from inorganic chemistry. From being a convinced advocate of Berzelius's views and a resolute opponent of splitting chemistry, Liebig became an opponent of Berzelius without, however, having been able to resolve the difficulties. On September 5, 1839, Liebig wrote to Berzelius:

You ask me whether, in fact, I accept radicals which contain chlorine. I am not uncertain about my answer. Permit me to state my view in another way. In inorganic chemistry, in fact, we do not know of any chloride in which the chlorine could not be replaced by oxygen, provided that as an electronegative body it takes the place of oxygen. Many carbon and other compounds of an organic nature have been found where the behaviour of chlorine proves...that it does not play the role of an electronegative body. You consider these compounds to be chlorides, which do *not* possess the properties of inorganic chlorides...We are quarrelling really about principles. You stand for the vindication of the existent ones, I am for the improvement, for the development of the same...We shall have laws of proportions for the combination of inorganic compounds. We shall obtain laws for certain series of organic compounds. What will be a rule for the one, will be an exception with the others. I am continually vexed by doubts which inorganic chemistry is not able to remove, we are unable to explain the simplest things; everything in organic chemistry is different (Berzelius & Liebig, 1893, p. 201).

The discussion between Berzelius and Liebig continued, but without results. The older chemist did not budge and the younger one, not really knowing the right answer, became very unsettled. The letter from Liebig to Berzelius of 26 April 1840 contains the following sentences:

I confess in advance, this is the expression of an insurmountable loathing and revulsion against what is at present happening in chemistry. The quarrel about the substitution theory has been pushed to extreme lengths...I am becoming completely dispassionate, cooler and more sensible...I have asked myself earnestly of what use are these arguments. Neither medicine nor physiology, nor industry can gain useful applications from them. Everything we have is too young and new that we may hope to develop laws which would last longer than a month...For 4 months I have been devoting myself to quite another aspect of the science. I have been studying organic chemistry in relation to its laws which in their present stage have a bearing on agriculture and physiology (Berzelius & Liebig, 1893, p. 210).

Liebig was referring to his studies which brought organic chemistry into close relation to plant and animal physiology.

PHYSIOLOGY

Let us now examine briefly how the relations of chemistry and physiology developed in this period. The hopes expressed by Reil in his article on the vital force matured only gradually. That the full significance of important chemical discoveries in relation to certain living processes was not fully realized, can be said both about the facts on photosynthesis which were brought to light successively by Priestley, Ingenhousz, Senebier and de Saussure, and on animal and human respiration which were discovered by Lavoisier, Laplace, Séguin and Crawford.

For many years botanists, plant physiologists and practical agriculturists continued to believe that plants derived their carbon from the humus and not from air. As for the work of Lavoisier which brought out the analogy between respiration and combustion, Berzelius maintained for many years that it was still entirely unknown how animals kept their temperature constant (Berzelius, 1831, p. 109).

We still know little about the reasons for this curious situation. It will be the task of the history of science to provide a substantiated answer to this and other questions. Meanwhile let us note that there was closer co-operation between physiology and chemistry in the field of nutrition. It may be argued that a comprehension of respiration of plants and animals presupposed a deeper understanding of various aspects of nutrition; this brings us up against the problem of the inner logic of the development of science. While this was undoubtedly a factor in bringing out an interest in the problems of nutrition, it was stimulated by social needs as well.

The great changes which were brought about in England by the Enclosure Acts profoundly influenced the diet of the labourer. This has been clearly pointed out by the Hammonds in their well-known book '*The village labourer 1760–1832*' in which they wrote:

Whereas his condition earlier in the century had been contrasted with that of continental peasants greatly to his advantage in respect of quantity and variety of food, he was suddenly brought down to barest necessities of life...The labourers were rapidly sinking in this respect to the condition that Young had described a generation earlier as the condition of the poor in France...All the budgets tell the same tale of impoverished diet accompanied by an overwhelming strain and actual deficit (Hammond & Hammond, 1936).

It is not difficult to understand that food shortage became an important social problem in England at the end of the eighteenth century. This also applied to many parts of Europe where scarcity of food and high prices prevailed early in the nineteenth century. In 1815 the Academy in Paris set up a commission which had to enquire into the possibility of replacing meat in the food of the poor by the gelatinous extract of bones. The commission was headed by the famous physiologist Magendie and Vauquelin was another member of it. Studies of food substitutes appeared in other countries too. Thus two scientists from Prague, Neumann and Steinmann, suggested in 1817 that Iceland Moss should be used as food substitute in place of cereals.

It may be asked whether the state of knowledge at that time made it possible to approach the subject of nutrition in a rational manner. By 1811 investigation of carbohydrates was fairly advanced. In that year Gay Lussac and Thénard for the first time determined accurately the elementary composition of cane sugar. About the same time, Chevreul began his famous investigation of fats. Also in 1811, Berthollet gave a new turn to the analysis of proteins when he measured the quantity of ammonia which was obtained from distillation of weighed amounts of meat and cheese.

In 1816 Magendie published a paper in which he examined the nutritive properties of substances lacking nitrogen. The real object of this investigation was to find out the origin of nitrogen which was known to be an abundant constituent of the animal body. On feeding dogs with sugar, olive oil or butter respectively and giving them distilled water to drink, Magendie discovered that they were unable to exist and after a time they died. He concluded by saying that the work showed, not positively but at least very probably, that nitrogen which was playing such an important role in the life of animals was largely of nutritional origin.

Besides Magendie there were others who enquired very closely into the chemistry and physiology of food. One able worker was the English physician William Prout, generally known as the person who proposed the theory that atomic weights of all elements were multiples of the atomic weight of hydrogen; he is largely forgotten as the man who for the first time proposed the division of foods into sugars, fats and proteins. This was suggested by him in a paper called 'On the ultimate composition of simple alimentary substances with some preliminary remarks on the analysis of organized bodies in general' which appeared in 1827 in the Philosophical Transactions. The relevant passage reads as follows: 'and by degrees I had come to the conclusion that the principal alimentary matters employed by man and the more perfect animals, might be reduced to three great classes, namely, the *saccharine*, the *oily*, and the *albuminous*' (Prout, 1827). Although he was originally more interested in problems of chemical proportions, Prout perceived that physiology and pathology could derive considerable help from chemistry. He tackled the problem of digestion and in 1824 was able to demonstrate the existence of hydrochloric acid in the stomach. His aim was to elucidate the changes which took place in the stomach and other organs when food passed through them. Systematic work on this was carried out by two German workers, Gmelin and Tiedemann, who published their findings in 1826. Among other things they confirmed Prout's contention that the gastric fluid contains hydrochloric acid.

One of the problems which faced investigators of digestion was how to explain its mechanism. Was digestion a purely chemical or was it a vital process? This is how the question was answered by Tiedemann and Gmelin: 'Even though the gastric fluid, as a result of its chemical composition, is the dissolving agent of both the simple and the composite foods and its action on food is a chemical one, digestion is still a vital process conditioned by the life of animals. That is, only the living stomach through its vital forces is able to secrete the dissolving gastric fluid' (1831).

The dilemma which Tiedemann and Gmelin placed before their readers in this clear way concerned not only digestion but any living process. The question was how to reconcile chemistry with life. It must be said that neither Tiedemann and Gmelin, nor the overwhelming majority of workers who came after them, were able to solve their difficulties in this respect. This then became the basis of the seemingly unending controversy between the mechanists and vitalists. The dispute brought fruitful results in the experimental sphere. Both sides, in striving to prove their case, were constrained continuously to seek new experimental evidence.

An important contribution to the study of digestion was made by Eberle, who by 1834 had shown that the presence of a stomach is not required for gastric digestion to take place. He demonstrated that gastric fluid is able to digest food outside the stomach. He also prepared a fluid which showed digestive properties, by digesting gastric mucosa with dilute hydrochloric acid.

Eberle's work had important repercussions. It was followed up in 1836 by Schwann's discovery that gastric fluid contains, in addition to hydrochloric acid, another digestive component which he named pepsin. Schwann tried to explain the mechanism of pepsin action by assuming that it was a catalytic or contact action, a view supported by the fact that an extremely small quantity of pepsin sufficed to dissolve a great quantity of albumen.

The term 'catalysis' had been introduced by Berzelius in 1835 to cover a wide range of chemical phenomena. Kirchhoff had shown earlier that starch could be decomposed by acids, and Davy and Döbereiner that platina was able to effect a transformation of methane or alcohol into other compounds.

It is thus fairly obvious that, in the second half of the thirties of the last century, there were growing signs of chemistry becoming in many ways useful, indeed indispensable, for a rational explanation of living processes. This was clearly understood by Purkinje and Schwann when they were working out the principles of the cell theory.

It is not widely known how Purkinje came to formulate his views on the analogy between the structural units of plants and animals. Purkinje, who was both a histologist and an experimental physiologist, undertook, together with his pupils at the University of Breslau, a systematic examination of the microscopic structure of animal tissues. In 1836 and 1837, following up the work of Eberle and Schwann, he conducted with Pappenheim some experiments on artificial digestion. It is very probable that Purkinje was the first to observe the inhibitory action of the gall on pepsin. At the same time he was examining the microscopical structure of the stomach. When addressing the annual gathering of German naturalists and medical men at Prague in 1837 on the glandular structure of the stomach and nature of digestion, Purkinje observed that the granular structure of the animal tissues showed analogies to the cellular structure found in plants. He called the substance of the granules 'enchyme', and two years later he used the term 'protoplasma' to describe its state in its embryonic stage. Despite this erroneous conception of the origin of the cells, Purkinje immediately realised the importance of studying metabolic changes which are taking place in the cells. He wrote:

The granular elementary form suggests again an analogy to the plant which is known to consist almost completely of granules or cells. The process of enchyme formation and disassimilation could be visualised from the way in which every little cell possesses its *vita propria* and prepares, from the general sap, its own specific contents and how, through its mediation, special materials are again discharged into specific sap reservoirs. In all, the present concept of the granular enchyme of the animal organism leads us again to a more exact study of plant physiology. So also the possibility thus opened up of preparing pure specific enchymes, will provide animal chemistry with ample material for further speedy advances (Purkyně, 1948).

It was not, however, Purkinje but Schwann who worked out the details of the cell theory in a systematic manner. His book *Mikroskopische Untersuchungen über die Uebereinstimmung in der Struktur und dem Wachsthum der Thiere und Pflanzen* appeared in 1839, and its importance was recognized immediately. The book contained three main sections, of which the last is of special interest to us because it raises directly the problem of the chemistry of the cell.

Schwann introduced his subject with a discussion on the relative merits of the vitalistic and mechanistic approaches to the elucidation of the fundamental basis of life. This was not unusual because all great physiologists of that period, to mention only Magendie, Müller and Purkinje, were perpetually vexed with this problem. Magendie was under the influence of mechanical materialism, Müller under German idealism, and Purkinje, whilst basically an idealist, was vacillating. Schwann, though in a way ambiguous, came in the end down on the side of materialism. This led him to an important formulation that 'the cause of nutrition and growth is not vested in the organism as a whole but in the individual parts, the cells.' This mechanist approach was then tempered with a fine observation: 'The manifestation of the force which lies in the cell depends on the conditions which can be provided for it only in connection with the whole' (Schwann, 1839, pp. 228–9). Schwann like Purkinje believed erroneously that cells came into existence in an elementary substance which he called, after Schleiden, cytoblasteme. He thought that the cells were able to draw materials from the cytoblasteme and change them chemically. 'Besides that', he wrote, 'all the parts of the cell can be changed chemically during its growth. We want to call the unknown cause of all these phenomena which we can sum up under the name of metabolic phenomena of the cells, *metabolic force'* (Schwann, 1839, p. 234). This was perhaps the first time that the term 'metabolism' was used in its modern biochemical sense. There is no space to dwell on other aspects of Schwann's work. For instance, he discussed in his book the colloidal state of living matter, though he did not employ the term 'colloidal'.

It may be added that, in spite of the fact that Schwann erred over the question of the origin of the cell, his concept of metabolism was not a product of pure speculation. We have to remember that Schwann, at about the same time but independently of Cagniard de Latour, was working on problems of fermentation. Already in 1818 Erxleben, a German industrial chemist who worked in Bohemia, suggested that yeast is a plant whose growth caused fermentation. Erxleben's view expressed in a small book *Über Güte und Stärke des Biers* remained apparently unnoticed. In 1835 Schwann published the results of his work on fermentation which showed that yeast was growing during alcoholic fermentation and that this was probably the cause of fermentation. In his book on the cell, Schwann referred to this work and pointed out in a footnote that the metabolic changes in the yeast cell represented the best known and the simplest example of a process which repeated itself in every living cell. Finally, he expressed the hope that the material he presented in the book would gain more favour for his theory of fermentation.

PHYSIOLOGICAL CHEMISTRY AND BIOCHEMISTRY

The further development of the study of chemistry of life confirmed Schwann's important observations but the historical development of the understanding of the role of ferments in the chemistry of the cell was not straightforward. Whilst the study of fermentation was gradually taken up in a systematic manner, the chemistry of the cell was neglected.

The obvious explanation for this might be that conditions were not ripe for studies which related fermentation with chemical changes in the cell, as brilliantly suggested by

Schwann. This explanation may be valid when we look at the problem today. But did it look so to the workers in the early forties of the last century?

On examination we find that the majority of investigators of those days, led by Berzelius and especially by Liebig, rejected completely the suggestion that fermentation had anything to do with life. On the contrary, Liebig defended for many years the view that fermentation was a purely chemical process closely related to putrefaction and decay of those organic bodies which had already ceased to live. Under these circumstances it is really not surprising that Schwann's suggestion was not taken up and that the study of fermentation developed at first independently of the cell theory.

Physiology—the science of life—was revolutionized in the forties on the basis of the cell theory and chemistry. It is interesting to note that Liebig apparently did not understand the significance of the cell theory for physiology while being fully aware of the importance of chemistry. Liebig began to enquire more deeply into the application of chemistry to agriculture, medicine and industry, when he was unable to see his way in the theoretical discussion on organic chemistry. Both physiology and chemistry experienced great changes at that time. The separation of inorganic and organic chemistry so deplored by Berzelius, and not understood by Liebig, was temporary because it enabled both branches to unite in the sixties on a higher level with the acceptance of the atomic theory. The character of organic chemistry had changed too: it stopped dealing with chemical processes in the organs of plants and animals, which became the task of physiological chemistry, and studied the properties of the carbon compounds.

Physiological chemistry thus inherited the problems which were previously investigated by organic chemistry. The significance of Liebig in this field does not lie so much in his scientific discoveries, which were frequently in error, but in that he understood the necessity of linking chemistry, particularly organic chemistry, to physiology, and in that he worked hard to apply plant and animal physiological chemistry directly to the service of agriculture, medicine and industry.

Mikuláš Teich

This he expressed very clearly in his famous book *Organic chemistry in its applications to agriculture and physiology*, first published in 1840 and written at the request of the British Association for the Advancement of Science. The dedication reads:

I have endeavoured to develop, in manner correspondent to the present state of science, the fundamental principles of chemistry in general, and the laws of organic chemistry in particular, in their applications to agriculture and physiology; to the causes of fermentation, decay, and putrefaction; to the vinous and acetous fermentations and to nitrification... perfect agriculture is the true foundation of all trade and industry—it is the foundation of the riches of states. But a rational system of agriculture cannot be formed without the application of scientific principles; for such a system must be based on an exact acquaintance with the means of nutrition of vegetables, and with the influence of soils and action of manure upon them. This knowledge we must seek from chemistry, which teaches the mode of investigating the composition and of studying the characters of the different substances from which plants derived their nourishment... (cf. Browne, 1944).

The companion book *Animal chemistry in its applications to physiology and pathology* appeared two years later.

From what has been said it becomes fairly clear that, by the beginning of the forties of the last century, various tendencies deriving from chemistry and physiology heralded a new phase in the development of the science of chemistry of life. These tendencies, which crystallized in the establishment of physiological chemistry, received a powerful stimulus from the needs of agriculture, medicine and industry.

The social impetus which drove Liebig and others to study problems relating to agriculture, medicine and industry derived from effects produced by the industrial revolution. Social relations of science take on different aspects in different periods of history. One of the distinguishing features of the period of the industrial revolution was that scientific knowledge became much more directly integrated with manufacture than before. That applied not only to industry but also to agriculture which had to provide for the ever increasing population of the towns. Briefly, fewer hands were available to feed more mouths:

The foundations of modern biochemistry

this called not only for the improvement of techniques in farming, but also for the application of scientific knowledge in agriculture. Agriculture was slowly being transformed into a branch of industry. This industrialization of agriculture was also reflected in the growth of the sugar beet, fermentation and other food industries. These then are the social roots, not only of Liebig's work on fermentation, but also of Pasteur's. Without understanding the social background, it is difficult to comprehend the full historical significance of the great dispute between these two scientists regarding the nature of fermentation. It would be just as wrong to see it in terms of a purely mechanistic–vitalistic controversy as to explain it in purely economic terms. However, both aspects contributed profoundly to the transformation of physiological chemistry into modern biochemistry, as also did the work of Voit and others on the value of foods.

It is perhaps fitting that one of the first to realize the significance of the Liebig–Pasteur discussion on the development of *biochemistry*, was a German wine-merchant named Moritz Traube who established a private laboratory of his own and published important experimental and theoretical papers on fermentation. Already in 1861 he realized that chemical processes and living bodies were mostly based on ferment actions, a fact which convinced him that an understanding of the chemistry of life is altogether impossible without a correct fermentation theory (Traube, 1899). There were others besides Traube interested in developing an understanding of fermentation. Thus Hoppe-Seyler had come to believe in the seventies that there is no need to differentiate between 'organized' or 'formed' ferments which were thought to function only inside yeast cells and the 'unorganized' ferments of the diastase or pepsin type (Hoppe-Seyler, 1877). In 1876 Kühne renamed the 'unorganized' ferments 'enzymes'. The experimental proof that there was no distinction between ferments and enzymes came in 1897 when Buchner showed that 'an extract of yeast cells was able to carry out fermentation'. This discovery made it possible to work out a correct theory of fermentation or enzyme action which, in the words of Traube, could be the only basis for biochemical research. At the same time, realization of this fact provided the key for studying chemical changes of the cells.

The circle was closed. But it is correct to describe the zig-zag development of the chemistry of life from Schwann to Buchner in terms of a circle when it would be more appropriate to speak of a spiral? Physiological chemistry was transformed into modern biochemistry and new scientific influences arising from structural organic chemistry, physical chemistry and other branches of science began to influence its course. New social stimuli based on the necessity of understanding deficiency and other diseases made themselves felt. But philosophically there was little advance and biochemical thinking remained on the whole a prisoner of the old mechanistic–vitalistic antithesis. It would be an interesting question to see whether this proved to be a retarding, an accelerating or an indifferent influence. However, this is outside the scope of this article.

To sum up, early scientific biochemistry formed an integral part of a still unified science of chemistry. The separation of organic from inorganic chemistry started simultaneously the separation of biochemistry from the general body of chemistry; at this stage it was hardly possible to distinguish between organic chemistry and biochemistry. The situation changed at the beginning of the forties of the last century when organic chemistry began to be applied systematically to agricultural, medical and industrial problems. Important social factors were driving this process forward: the rapid development of the industrial revolution created new problems in food production, and chemistry began to penetrate into the sugar beet and fermentation industries. As a result of all this, the need for an understanding of chemical changes in living matter under normal physiological and pathological conditions increased tremendously. Philosophical views played an important role as well.

The separation of biochemistry from physiology started somewhere about 1870 to 1880. An interesting document about this exists in the controversy between Pflüger and Hoppe-Seyler when in 1877–8 the *Zeitschrift für physiologische Chemie* was established. Pflüger, the editor of the *Archiv für Physiologie*

The foundations of modern biochemistry

deprecated in no uncertain terms the establishment of the new journal. He believed that physiological chemistry formed an inseparable part of a unified science of life, i.e. physiology. The new journal still carried the name associating it with physiological chemistry but Hoppe-Seyler in his preface significantly pointed out: 'Biochemistry ... from its first natural and necessary analytical beginnings has grown into a science ...' (Hoppe-Seyler, 1877).

REFERENCES

Berzelius, J. J. (1827). *Lehrbuch der Chemie, vol. III*; (1831) *vol. IV.*
Berzelius, J. J. & Liebig, J. (1893). *Ihre Briefe von 1831–45.*
Berzelius, J. J. & Wöhler, F. (1901). *Briefwechsel zwischen Berzelius und Wöhler.*
Bodenheimer, F. S. (1958). *History of biology*; Goodfield, G. J. (1960). *The growth of scientific physiology.*
Browne, C. A. (1944). *A source book of agricultural chemistry/Chronicka Botanicka*, **8**, p. 264.
Hammond, J. L. & Hammond, Barbara, (1936). *The village labourer 1760–1832*, p. 87.
Hoppe-Seyler, F. (1877). *Physiologische Chemie*, pp. 114.
Hoppe-Seyler, F. (1877/78). *Zeitschrift f. Phys. Chemie*, **1**, 1.
McKie, D. (1944). Wöhler's 'synthetic' urea and the rejection of vitalism: a chemical legend. *Nature*, **153**, 610.
Magendie, F. (1816). *Ann. Chim. Phys.* **3**, 66–77.
Müller, J. (1835). *Handbuch der Physiologie des Menschen*, **1**, 8.
Needham, J. (1956). *Science and civilisation in China*. Cambridge, Vol. II, p. 302.
Neumann, K. A. & Steinmann, J. (1817). *Anleitung zur Benutzung des islaendischen Mooses zu gesunden und kraeftigen Nahrungsmitteln fuer Menschen, bei Mangel des Getreides.*
Prout, W. (1827). *Philosophical Transactions*, p. 337.
Purkyně, J. E. (Purkinjè) (1948). *Opera Selecta*, p. 110.
Reil, J. C. (1796). *Arch. Physiol. I.*
Schwann, Th. (1836). *Ann. d. Pharm.* **20**, 28–34.
Schwann, Th. (1839). *Mikroskopische Untersuchungen über die Uebereinstimmung in der Struktur und dem Wachsthum der Thiere und Pflanzen.*
Tiedemann, F. & Gmelin, L. (1831). *Die Verdauung nach Versuchen,* **1**, 336.
Traube, M. (1899). *Gesammelte Abhandlungen*, p. 172.

8

SOME LONE PIONEERS OF BIOCHEMISTRY IN THE NINETEENTH CENTURY

by Sir Rudolph Peters

Before dealing with biochemistry in Britain during this period something should be said about the continental background of the subject. For earlier work than the last century, it is worth perusing Michael Foster's *History of physiology* (1901), to get an idea how long there has been a swing between the theories of vitalism and mechanism in some form. On the continent, there was definite biochemical activity; there does not seem to have been quite the scorn of 'Slime' chemistry, as it was termed, which was to be found in Britain at that time. A marked exception must have been Professor Cohen of Leeds, who was responsible for the training of distinguished biochemists like H. D. Dakin, H. W. Dudley, H. S. Raper and H. Raistrick. Thus we can point to the synthesis of urea by Wöhler in 1827, to all the work of Claude Bernard on sugars and related substances, and to his views on the need of studying pathology at a chemical level; much of what he wrote is modern even today. We may instance the following passage from his works quoted by Professor Florkin. 'Les propriétés vitales ne sont en realité que les propriétés physico-chimiques de la matière organisée.' 'Une grande erreur consiste à croire que les phénomènes physico-chimiques de l'organisme sont identiques à ceux qui se passent en dehors, et d'avoir voulu expliquer ces phénomènes dans le corps par des agents qu'on avait consistés au dehors. C'est dans cette erreur que sont tombés certains chimistes, qui raisonnent du laboratoire à l'organisme, tandis qu'il faut raisonner de l'organisme au laboratoire.' Claude Bernard understood very clearly that the fact of its biological organization imposed certain differences upon a cell system otherwise behaving chemically.

On the continent there was the famous controversy about fermentation between Liebig and Pasteur, which you can read about in the introduction to Harden's *Alcoholic fermentation*; it is not clear why it was that none suspected the presence of a living organism until three workers independently reached the idea that this was the case in fermentation in 1837, incidentally the year in which Queen Victoria came to the throne. The strength of the feeling against their views is to be seen by the skit published in *Annalen*, so well described by Harden in the following passage.

To the scorn of Berzelius was soon added the sarcasm of Wöhler and Liebig (1839). Stimulated in part by the publication of the three authors already mentioned, and in part by the report of Turpin (1839), who at the request of the Academy of Sciences had satisfied himself by observation of the accuracy of Cagniard-Latour's conclusions, Wöhler prepared an elaborate skit on the subject, which he sent to Liebig, to whom it appealed so strongly that he added some touches of his own and published it in the *Annalen* following immediately upon a translation of Turpin's paper. Yeast was here described with a considerable degree of anatomical realism as consisting of eggs, which developed into minute animals, shaped like a distilling apparatus, by which the sugar was taken in as food and digested into carbonic acid and alcohol, which were separately excreted, the whole process being easily followed under the microscope.

Even if we realize that Liebig was trying to save the problem for chemical investigation, it should be a warning to all to be careful to look well into new findings. In a way there is a real dilemma for the scientist. He must necessarily carry criticism to the limit in examination of experimental facts. At the same time, a degree of open-mindedness may be required, which almost runs counter to his training.

In the controversy which developed, the papers of Liebig are theoretical to a fault, whereas those of Pasteur are clear and concise, 'if he did not go beyond his facts, no fermentation without life'. M. Traube did in fact suggest the right solution though it seems to have received little attention; and it was not until the Buchners in 1897 showed that a juice expressed from yeast under high pressure would carry out fermentation that it became generally accepted that a living organism made the enzymes which carried out the process.

With all this excitement on the continent of Europe, it comes about that when we ask the question, what was done in Britain previous to 1900; the answer is very little as a continuous effort. However certain individuals did in fact make pioneer observations of considerable importance. What is especially interesting is that most of these pioneers had been trained in medicine. They had a desire to help their patients and became genuinely curious about things which turned up in their practice.

As a beginning, we may mention W. H. Wollaston (1766–1828, plate 1). In 1797, he found uric acid in one of the chalky deposits in gout; and in 1810 he found the amino-acid cystine in a kidney stone. The observation had to wait until 1837 for the proof that this contained sulphur. Its origin in the body was not known until eighty years later, when it was shown by Mörner to be a constituent of normal skin. In fact the preparation of the beautiful hexagonal crystals of cystine from hair became a class experiment.

Wollaston had an unusual career. Born at East Dereham in Norfolk, he entered Gonville and Caius College in 1782, was a Scholar and took his M.D. degree in 1793, after becoming Senior Fellow in 1787. He was even a Tancred Student in Greek and Hebrew. In 1794, he became a Fellow of the Royal Society and in 1795 a Fellow of the Royal College of Physicians. It is to be noted that early in his career he practised medicine in Huntingdon and in Bury St Edmunds; but he retired from this about 1800, and embarked upon a scientific career of pure metallurgical research in the field of the noble metals, using for this a private laboratory of his own. He is now mainly known for this work, for instance the separation of palladium from platinum. In 1802 he reached the distinction of the award of the Copley Medal of the Royal Society, their highest award. It is clear that his biochemical observations sprang from his early medical interest; but that he had research instincts.

A little later, in 1835, Andrew Buchanan (1798–1882) in Glasgow (plate 2) studied the coagulation of the blood; in his back consulting room, as it were, in the midst of a very busy life, he found that blood coagulation was a process requiring two components. His fundamental observation was that hydrocoele fluid would remain stable for long periods without clotting, but that it would clot quickly if the washings of a blood clot were added to it. This theme was elaborated in several papers, and as Gowland Hopkins used to point out, he got almost as far as the whole school of Schmidt of Dorpat, who rediscovered some of his essential facts. Andrew Buchanan was evidently much respected in Glasgow, passing through the stage of being Professor of Physiology, and finally becoming Professor of Surgery, where he established a reputation in surgery by developing an instrument for the operation of lithotomy in 1845. Born in Glasgow, his father was a merchant and his mother came from Edinburgh. He was educated at the Glasgow Grammar School and entered the University about 1811, taking his M.D. in 1822. Later he studied in Paris, visited Italy, and founded the *Glasgow Medical Journal*. In one of his obituaries, the following occurs:

The tall, slightly stooping figure, the large venerable, tremulous head; the thoughtful, refined and benevolent face, which beamed with kindness, and the cordial, happy recognition with which he never failed to salute a friend. His unfailing politeness and courtesy, his genuine honesty of purpose, and the generosity and fairness which marked his intercourse with his professional brethren, made him beloved and trusted by all who came in contact with him. To intellectual powers of a high order he united the simplicity and guilelessness of a child.

In 1850, Dr H. Bence-Jones, F.R.S. (1814–73, plate 3), Fellow of the Royal Society and of the Royal College of Physicians published his lectures given in St George's Hospital with an expressed gratitude to Dr Prout who 'first established connexion between chemistry and medical practice'. As pointed out by Professor McIlwain (1958), several other books on urine analysis appeared from 1854 to 1865. In the book by Bence Jones, not only was the detail well given for practitioners of urinary analysis; but the chapter in this book contains an account of the method of detecting an unusual protein in the urine (p. 108). This was a new albuminous substance which precipitated from the urine with alcohol, and had the remarkable property of dissolving in boiling water, and coming out on cooling. It could be missed by the ordinary test with nitric acid, and it came from a patient with myeloma, cancer of the marrow. The peculiar physicochemical properties of this protein continued to be of interest to biochemists, and in 1911, F. Gowland Hopkins with H. Savory published a study of three cases of Bence-Jones proteinuria. I remember that Hopkins felt that these unusual solubilities would throw light on colloids generally. The work, which must have been extensive, did not do this, but Hopkins' instinct was evidently justified; because there is increasing study at the present time of the peptide chains of this unique class of proteins in the attempt to unravel the mystery of immune proteins.

In personal details of the life of Dr Bence-Jones, it is to be noted that he also came from East Anglia, from Thorrington Hall, Beccles, Suffolk. Educated at Trinity College, Cambridge, he became an M.D. in 1849, and eventually a full physician in St George's Hospital, London, where he spent his main life. At one time he was a Secretary of the Royal Institution, and is described as 'of a genial temperament'.

From a recent annotation (Clamp, 1967), it seems that the actual discoverer of the Bence-Jones protein was William MacIntyre, who was then fifty-two years old and Physician to the Metropolitan Convalescent Institution. He was called from Harley Street in 1845, to see a patient under the care of Dr Thomas Watson, and made many significant observations on the disease, as well as accurate experiments on the behaviour of the unusual protein in the urine. A specimen of this was sent to Dr Bence-Jones for further observation. Nevertheless Bence-Jones must have been a scientist of considerable ability. His capacity was also shown in another study, described in a lecture to the Royal Institution in 1966. Wishing to study the way in which drugs were distributed in the tissues after administration, he took advantage of the fact that quinine fluoresced blue in the ultraviolet, using this as a marker in modern terms. To his surprise he found that most animal

Plate 1. Portrait of W. H. Wollaston
(By kind permission of the Master and Fellows of Gonville and Caius College)

facing page 196

Plate 2. Portrait of Andrew Buchanan
(Reproduced by courtesy of the University of Glasgow)

Plate 3. Portrait of Dr H. Bence-Jones (1814–73)
(Reproduced by kind permission of the Royal Society of Medicine from an
engraving in their possession)

Plate 4. Portrait of Alfred Baring Garrod (1819–1907)
(From an original photograph in the Wellcome Historical Medical Museum and
Library. By courtesy of the Wellcome Trustees)

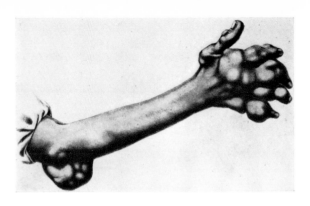

Plate 5. Gout, as seen by A. B. Garrod

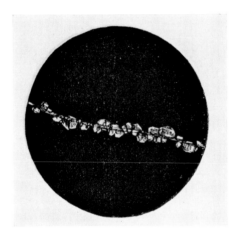

Plate 6. Uric acid test

Plate 7. Portrait of Archibald E. Garrod

Plate 8. Dr Sheridan Lea, Sc.D. F.R.S. (1853–1915)
(By kind permission of the Master and Fellows
of Gonville and Caius College)

tissues gave a blue fluorescence by themselves, and termed the substance concerned 'animal quinoidine', which he thought had many of the chemical properties of quinine itself. It was an interesting pioneer research; we now know as is so often the case in biochemistry that there must be several substances with these properties, one of which is formed by thiamine by closure of the ring, being called thiochrome, and much more fluorescent than quinine itself.

We can now turn our attention to 1860 and to gout. This was the year in which Sir Alfred Baring Garrod M.D., F.R.S., F.R.C.P. (1819–1907, plate 4) published the first edition of his *Treatise on gout and rheumatic gout* (rheumatoid arthritis). An examination of this work, of which I know best the third edition in 1876, is a revelation to anyone of the present day of the savage character of gout attacks in the past. Sydenham some 200 years earlier had pointed out that gout attacked the rich men rather than the poor. Pictures given in Garrod's book (plate 5) show that in more recent times, though gout still happens, it rarely seems to reach the same crippling severity. Garrod studied carefully the amount of uric acid in the blood of his patients. In the course of this he showed a remarkable biochemical capacity, by introducing a micro-test for uric acid in the serum from blood, which we can regard as a precursor of modern chromatographic procedures. His description of the method is very detailed. The whole was carried out in a flattened glass dish, about 3 inches in diameter. After adding acetic acid and mixing, one or two ultimate fibres, about an inch in length, were introduced 'from a piece of unwashed huckaback or other linen fabric, which should be depressed by means of a small rod, as a probe or the point of a pencil'. The dish was set aside for two or three days in a cool place, like a mantelpiece. When there is much uric acid present, it crystallizes on the fibre 'assuming forms not unlike that presented by sugar-candy upon a string' (plate 6). It was a fine micro-biochemical estimation, and his son told me that he remembered these flat dishes standing on his father's mantelpiece. Alfred Garrod was Professor of Therapeutics at King's College Hospital, London, to which he was also consulting physician.

It is interesting to note that he also was a native of East Anglia; he was born in Ipswich and educated at the Ipswich Grammar School, where Sir Charles Sherrington was at one time a pupil, as also Wolsey. Garrod took his degree in medicine in London, and was at University College before going to King's College Hospital. One of his great achievements was a book on Materia Medica.

It was a family with biochemistry in their chromosomes, because the fourth son of Alfred Garrod was Sir Archibald E. Garrod (1857–1936, plate 7) the physician at St Bartholomew's Hospital, who succeeded Sir William Osler as Regius Professor of Medicine in 1920, and whom I had the privilege of knowing rather closely while he was at Oxford. His elder brother who died at the early age of thirty-four, was already a Fowlerian Professor at the Royal Institution. Archibald E. Garrod developed the conception of Inborn Errors of Metabolism, based upon research upon certain rare diseases, especially alcaptonuria, in which the patient might excrete black urine from birth, and upon haematoporphyria congenita, associated with a port-wine-coloured urine. He rightly and very early recognised that one should not take a 'block' view of metabolism, and advanced the theory that in each of these cases, there was an enzyme missing, due to a genetic error. In this, he really proved that a disease could be caused by a deficiency, a view which even in the 1920s in connexion with vitamin deficiencies Gowland Hopkins was trying to bring home to medical men who were still imbued with a 'bacterial' view of all disease. As is now well-known Garrod's work was so significantly developed by G. W. Beadle and others, that it now forms part of the lore of molecular biology.

It is natural to associate Archibald E. Garrod with Sir F. Gowland Hopkins, because they worked together on urobilin in urine when Hopkins was at Guy's Hospital in London. During this period, until he went to Cambridge in 1899, I should also count Hopkins among the lone pioneers. Not only was there this work with Garrod; but there was the fascinating work on butterfly pigments, which he considered to be derivatives of uric acid. Since then however, this work has attained much more interest from the fact that the parent substance in these wings was later proved by Wieland not to be uric acid, but a pteridine. The pteridines have become enormously important as constituents of folic acid. *Inter alia,* they are much concerned with transfer of methyl groups in biochemical processes.

Two other observers made important discoveries in Oxford in 1898, J. S. Haldane (senior) published research upon the formation of carbon monoxide haemoglobin and its dissociation in high concentrations of oxygen. W. Ramsden, who became Professor of Biochemistry in Liverpool in succession to Professor Benjamin Moore, published his work on the denaturation of proteins by high concentrations of urea. Coagulated proteins could be brought into solution, and even a dead frog placed in saturated urea solution becomes transparent and falls to pieces in a few hours.

After 1900, the history of biochemistry in Great Britain is more continuous, and some of this is dealt with in my Hopkins Memorial Lecture, 1959. But there is one other name which must be mentioned in an account of this kind, namely Thudichum. Much valuable information about him has been given already by Dr K. C. Dixon in his lecture on the brain. To this I can only add a few comments in extension of what he has said, most of which I owe to a personal communication from Sir Henry Dale, O.M., F.R.S. He exemplifies so well two points which I have been trying to make in this lecture. One is the quasi saint-like devotion to science of some of these pioneers. Thudichum writes in his book:

Premiums in the shape of sensational discoveries may be hoped for, but cannot be assured even to the greatest genuis. But what has to penetrate, relative to this question, more completely into the consciousness of pathologists, is this, that to understand zymoses, to be able to counteract them by rational, as distinguished from empirical or accidentally discovered means, is only possible by the aid of a complete knowledge of the chemical constitution of all the tissues, organs and juices of the body, and of all their possible products.

It is good to know that the specimens from brain representing his work now reside in London owing to the efforts of Dr Otto Rosenheim.

Sir Rudolph Peters

The further point to be learnt from the career of this man as in the case of my earlier quotation from Harden's book, is the lengths to which scientific controversy could drive people in the past. I think that there can be unjustified scepticism even today, but it has not the venom of the past. Sir Henry Dale has written that Thudichum was assailed, denounced and ridiculed by leaders in the field of research both in Germany and England; for example, Gamgee said that one might as well supply a formula for bread and butter; Hoppe-Seyler, 'obviously false'; Maly, 'Mr Thudichum's dilettantism goes on, an epidemic disease for physiological chemistry'—and Sir Henry has described how this persisted in the minds of folk even until 1908—and all the time, as Rosenheim showed so well, Thudichum was right. It was this clamour of criticism which in the end led to his grant being discontinued, and so to his return to a nose, ear and throat practice. In spite of it all he continued his work with an assistant in the greenhouse of a Victorian house, half way up the stairs. Sir Henry has recorded a personal memory of a dramatic occasion.

Professor Gamgee attended a meeting of the Physiological Society in 1909, leaving his retirement in Switzerland. The Professor in whose Department this meeting was held, the late Dr A. D. Waller, was so much impressed by the return of the famous veteran, that he departed from the usual procedure and, amid general applause, invited Dr Gamgee to act as Chairman for this occasion. It happened that there was again a communication from Otto Rosenheim on the programme of this meeting; it dealt with the non-contentious subject of the lipids of the adrenal cortex. When Rosenheim had finished his exposition of it, Dr Gamgee rose and, with an old-world courtesy, congratulated him on the interest and importance of the work on which, he said, it was now his pleasant duty to call for discussion, etc. We seemed to be safely past the danger signal; but, while our veteran colleague was still standing, the old memories began visibly to stir in him, so that the colour of his complexion began to present a deepening contrast to his shock of white hair. 'But before I sit down, he exclaimed, I cannot let this opportunity pass without expressing my conviction that every word that Dr Rosenheim has ever published on the subject of Protagon is absolutely untrue and completely unjustified',—and when at last, he had sunk trembling into his seat, there was a long and painful silence, until he had pulled himself together, so as to be able to call upon the next contributor. And it was not long after this meeting that Dr Gamgee died. *

Though Dr Arthur Gamgee, M.D., F.R.C.P., F.R.S. (1841–1909) was not a lone pioneer in the sense of many mentioned here in this chapter, he was an important figure in our subject. At one time he was Professor of Physiology in Owen's College, Manchester. He published observations on optical properties of haemoglobin and turacin, and studied the action of nitrites on blood, developing the use of amyl nitrite in medicine.

Another interesting example of a man who made very significant observations on porphyrin pigments in muscle tissue, so-called histohaematin and myohaematin, is that of Dr C. A. MacMunn (1852–1911). He was educated at Trinity College, Dublin. During a busy medical practice his observations again were mostly made in a small laboratory in the hay-loft over his stables, in the afternoon before his afternoon round. His work was spectroscopic and in this field he was a recognized expert. His observations on muscle pigments however contained certain contradictions which could not be resolved at the time and which led to failure of acceptance by persons like Sir Ray Lankester and Professor Hoppe-Seyler. His disappointment over this led him to much worry and may have contributed to his death in 1911. Later in the 1920s, Professor D. Keilin rediscovered some of his facts, and made much progress in the field with his researches on cytochrome. A very good account of MacMunn's work is to be found in his book.

In the spectroscopic field, it may be noted finally that the first spectrum of chlorophyll was determined by Sir David Brewster (1781–1868), an F.R.S. and Copley Medallist.

Another physiological chemist of the last century who worked on the action of ferments should be mentioned here, because he made considerable contributions to the teaching of the subject. This was A. Sheridan Lea, Sc.D., F.R.S. (1853–

---

* For some further remarks, see p. 202.

1915, plate 8). He wrote the chapters on chemical physiology in the later edition of Michael Foster's *Physiology*. Lea was born in New York, became an undergraduate at Trinity College, Cambridge, and was finally Fellow and Bursar of Gonville and Caius College.

To conclude, I hope that these somewhat brief remarks about researches carried out in Britain by lone pioneers will suffice to show that there were dedicated persons, who even in the lack of organized biochemistry were able to think about some of the problems and to make contributions of real value. Most of them started with a training in medicine, and research was a hobby rather than a profession. As I have pointed out elsewhere, with some marked exceptions, biochemistry in Britain developed within the aegis of physiology rather than in that of chemistry.

APPENDIX

Sir Henry Dale added the following remarks.

Thudichum had kept his flag flying to the end. He seems to have lived a full social life in the circle of his friends. He was evidently a connoisseur of wines, as well as a keen investigator of their chemistry. When he died he had little to leave for the support of his widow and his four daughters, who faced the situation with courage, and made themselves self-supporting, in a manner much less common then than in the England of today. With advancing years, however, the strain became eventually too heavy. Dr Rosenheim had made touch with them, and some of us, when he had made the situation clear, found great satisfaction in organizing a petition to the Prime Minister of the day, which led to the granting of a pension to these ladies from the Civil List, in recognition of the distinguished services which their father had rendered to science. They, in turn, presented, through Dr Rosenheim, the most interesting collection of pure chemical substances, all being original products of their father's great researches, which now, in their show-case, are displayed in the Library of the National Institute for Medical Research. A photograph of this collection was sent for the information of those who were able to attend the meeting held to commemorate Thudichum and his works.

Pioneers of biochemistry in the nineteenth century

Acknowledgements

I should like to thank the Librarian of the Cambridge Philosophical Society, the Librarian of the Royal Society of Medicine and the Master of Gonville and Caius College, Cambridge, for their help in the preparation of this lecture.

REFERENCES

Brewster, David. (1834). *Trans. Roy. Soc. (Edinburgh)*, **12**, 538.

Buchanan, A. Obituary in *Glasgow Medical Journal* (1914), **81**, 432.

Clamp, J. R. (1967). *Lancet*, p. 1354.

Dale, H. H. (Personal communication).

Florkin, M. (1967). Quoted from 'Claude Bernard et les débuts de la biochimie' in *Philosophie et méthodologie scientifiques de Claude Bernard*. Paris, Masson, 1967, pp. 57–63.

Foster, (Sir) Michael. (1901). *History of physiology*. Cambridge University Press.

Garrod, A. B. (1876). *Treatise on gout and rheumatic gout*. Green and Co., London.

Garrod, A. E. (1923). *Inborn errors of metabolism*. Second edition. Henry Frowde and Hodder and Stoughton, London. See also W. E. Knox. (1967). *Genetics*, **56**, 1.

Harden, A. (1923). *Alcoholic fermentation*. Longmans, Green and Co., p. 8.

Jones, H. B. (1850). *On animal chemistry in its application to stomach and renal diseases*. John Churchill, London.

McIlwain, H. (1958). Thudichum and the medical chemistry of the Eighteen Sixties to Eighties. *Proc. Roy. Soc. Med.* **51**, 127.

Peters, R. A. (1959). Hopkins Memorial Lectures. *Biochem. J.* **71**, 1.

INDEX

Abel, J. J., 135
acetic acid, synthetic, 176
acetyl choline, 110
acromegaly, 147–8
Adams, C. W. M., 103, 104, 113
Adams, Robert, 77
Addison, Thomas, 131–3, 149
Addison's disease, 131–2, 136, 146
adenosine triphosphate, 30, 112, 113
adrenal glands, 128–9, 130, 131, 133–6, 151; cortex of, 136, 141, 145–7, 148, 149
adrenaline, 110, 135–6, 141
adrenochrome, 111
Adrian, E. D., 97, 115–16
air: carbon dioxide in, 6; composition of, 10–12, micro-organisms from, 42, 43, 45; plant nourishment from, 9
Albers, R. W., 110
Albertus Magnus, xxiv
alcaptonuria, 198
alchemy, xiv, xxv–xxvi, 7; of Arabs, xiv, xviii–xix, xxv; of Chinese, xvii–xviii, xxi; of Indians, xxii–xxiii
Alcmaeon of Croton, 61
alcohol, distillation of, xiv
alcoholic fermentation, 47–8, 49, 50
aldosterone, 147
Aldrich, T. B., 135
Allen, F. M., 138
b-aminobenzoic acid, 166
amylases, 17
amyloid bodies, 88
anabolism, equilibrium between katabolism and, 82–3
anaemia, Addisonian or pernicious, 132, 133, 166
anaerobiosis, 41, 50, 55
Anaxilaus of Larissa, xvii
Anderson, J. P., 110
androsterone, 145
aneurin, 167
Anguiano, G., 114
Animal Chemistry, Society for the Improvement of, xx n.
animal-protein dietary factor, 166
'animal spirits', x, 62, 64

Ansbacher, S., 166
anthrax, bacilli of, 53–5
antisepsis, 52
Appert, P., 42, 43
Aristotle, x, xxiv, 5, 7, 10
Arndt, R., 94
Arthus, M., 22
Aschheim, S., 144, 148
Aschner, B., 148
asepsis, 55
Ashford, C. A., 113
atomic theories, viii, xii, 4
axoplasm, flow of, 101, 112

Bach, A., 32–3
Bacon, Roger, xvi
bacteria: discovery of, 39, 40, 41; fermentations by, 48, 50; photosynthetic, 13; pure cultures of, 49, 53
Bailey, C. V., 143
Bang, I., 143
Banting, F. G., 139, 141–4
Barendrecht, 22
Bargmann, W. von, 113
Barnes, C. R., 1
Barr, M. L., 100
Bartholinus, C., 131
Bassi, A., 52
Bastian, H. C., 46
Batelli, F., 30
Baum, H. M., 166
Baumann, E., 137
Bayliss, W. M., 125, 126, 141
Beadle, G. W., 58, 198
Becher, J. J., 10
Becker, J. E., 163
beer, maladies of, 50–1
Bell, Charles, 74, 150
Bence Jones, H., 195–6
Benedictine, xv
Berger, Hans, 115
beri-beri, 157, 158, 159–60, 163, 167
Berkeley, M. J., 52
Bernard, Claude, 127, 133, 150, 192
Berthelot, P. E. M., 21
Berthold, A. A., 130

Index

Index

Index

Index

Hoffer, A., 111
Hokan Chi, xvii
Holmes, E. G., 109
Holmes, Gordon, 96
Holst, A., 158, 160
Hopkins, F. G., xiii, xxvii, 35, 133, 156, 158, 160, 161, 162, 164, 195, 196, 198
Hoppe-Seyler, F., 32, 84, 89, 138, 189, 190, 200, 201
hormones (animal), 125–55
Horsley, V. A. H., 137
Houssay, B. A., 149
humours, doctrine of, xii–xiii
Huntsman, M. E., 165
Hsün Ch'ing, x
Hu-Ssu-Hui (Hoshoi), xi
Hydén, J., 99, 101, 117 n.
hydrochloric acid, in stomach, 182
hydrogen, carriers of, 29, 34–5
hydrogen peroxide, 32, 33–4
5-hydroxytryptamine, 110–11
hypoglycaemia, 143
hypothalamus, 151

iatro-chemistry, xx, xxvi; Chinese, xx–xxii
impulse-propagating substances, 111–12
'indophenol oxidase', 108–9
Ingenhousz, J., 11, 12, 180
industrial revolution: and applied science, 188–9; and nutrition, 181
inositol, 166
insulin, 106, 139–40, 141–4; pituitary and, 149–50
iodine, in thyroid, 136–7
ion exchange, 146
iron, in enzymes, 31, 34

Jābirian Corpus, xiv, xviii ff., xxv
Jacob, L., 161
de Jager, 22
Joblot, L., 42
Johnson, A. C., 103

Kabat, H., 110
Kant, I., 115, 119
Kastle, J. H., 31
katabolism, equilibrium between anabolism and, 82–3
Katsoyannis, P. G., 144
Katz, J., 117 n.
Keilin, D., 34, 35, 108, 201

Kendall, E. C., 141, 147
kephalins, 87, 102
kerasin, 87
Keynes, R. D., 113
kinetics, of enzyme action, 27–8
King, T. Wilkinson, 130
Kingsbury, B. F., 37
Kirchoff, G. S. G., 184
Klein, J. R., 114
Klenk, E., 102
Knight, B. C. J. G., 57
Ko Hung (Pao P'u Tzu), xviii
Koch, Robert, 53, 54, 56
Kolbe, A. W. H., 176
Kölliker, R. A. von, 79, 96
Kramer, J. G. H., 158
krasis (balanced admixture), xiii, xiv
Kuhn, R., 163
Kühne, W., 21, 189
Kunitz, M., 24
Kützing, F., 47–8

lactic acid, in brain, 87, 107, 109
lactic fermentation, 48–9
Lancaster, J., 158
Lankester, Ray, 201
Langerhans, P., 142
Laplace, P. S. de, 180
Lashley, K. S., 115
Lavoisier, A. L., viii, xix, 10, 12, 31, 172, 178, 180–1
Lea, Sheridan, 85, 99, 201
Leeuwenhoek, A. van, 38–41, 58
lecithins, 84, 102
lei (categories), xxiii
Leiner, K. Y., 110
Lepage, L., 125
Lepkovsky, S., 165
Li Cho-Hao, 149
Li Shao-Chin, xvii, xix
Libavius, Andreas, xix
Liebig, J. von, xi, 19–20, 31, 85, 179–80, 187–9, 193
Liebreich, O., 84, 89
light, and plants, 9, 11
Likely, G. D., 166
Lillie, R. D., 163
Lind, J., xxvii, 158
Lindsay, H. A., 100
Lineweaver–Burk plot, 28
lipids: of brain, 84, 87, 88, 118; of grey matter, 102–3; intraneuronal, 96–7; of white matter, 103–4

209

Index

Index

neurones: anoxia of, 105, 106; cerebral, as independent units, 92–4; chemistry of, 94–7; macromolecular fabric of, 98–104, 116–17; as metabolic units, 79–83; *see also* nerves
neurosecretion, by hypothalamus, 113
nicotinamide, 164
nicotinamide-adenine dinucleotides (NAD, NADP), 36
nicotinic acid, 164
Nielson, E., 166
Nissl, F., 94
Nissl substance, 94–6, 98, 99
nitrogen, in nutrition, 182
Noback, C. R., 81
Noeggerath, C. T., 161
noradrenaline, 110, 135
Norris, L. C., 165
Northrop, J. H., 24
nucleic acid: cytoplasmic, 95, 99; in neurones, 98, 99, 118
nucleoprotein, 95, 98
nutrition: of bacteria, 57, 58; chemistry and physiology of, 180–1; of nerves, 69, 81, 82, 95, 101–2

Oliver, George, 134, 147, 148
Olsen, N. S., 114
Olympiodorus, xviii
Opie, E. L., 139
Ord, W. M., 137
Osmond, H., 111
'outer' and 'inner' elixirs, xxi
oxidations, 30–7; by brain, 105, 114
oxygen, 12; in respiration, 31; supply of, for brain and nerve, 78, 104–10, 113; utilization of, by cells, 16, 35
oxygenase, 33
oxytocin, 149
ozone, 31–2

pancreas: and diabetes, 137–40; insulin from, 141–4; secretin and, 126
pantothenic acid, 165
Pappenheim, S. M., 184
Paracelsians, x, xi, xii, xx, xxiii
Paracelsus, xix ff., 136
parathyroid glands, 141
Parmenides of Elea, 4
Pasteur, L., xi, 19, 20, 43–6, 48–51, 53, 56, 189, 193
Pasteur effect, 107
Pavcek, P. L., 166

Payen, A., 17
Pekelharing, C. A., 158, 161
pellagra, 158, 163, 164, 167
Pelletier, B., 13
P'êng Ssu, xxi
pepsin, 17, 22, 183, 184
peroxidase, 31, 32, 33
peroxides, 32
Persoz, J. F., 17
Peters, R. A., 108
Pfeffer, W., 1
Pfiffner, J. J., 145–7
Pflüger, E., 150, 190–1
Phillipeaux, M., 134
Phillips, D. C., 26
Philon of Byzantium, 6
phlogiston theory, 10, 11, 12, 31
phosphate bonds, high-energy, 112, 113
phosphatides (phospholipids), 84, 87, 88, 118
phosphocreatine, in brain, 112
phosphoprotein, in nerve, 112
photosynthesis, 1–14
phrenosin, 87
physiological chemistry, and biochemistry, 186–91
physiology, relations of chemistry and, 180–6
pineal gland, 64, 65, 67
pituitary gland, 129, 135, 141, 144, 147–50, 151
plants, 1–2; nutrition of, 7–8; and water 3, 4, 7
Plato, xii, 4, 5
Pliny the elder, 6
pneuma, ix–xi
polypeptide hormones, 141, 149
Pope, Alexander, 119
Popielski, L., 125, 126
Posener, K., 106
potassium, in nerve metabolism, 111, 113, 114
Pouchet, F. A., 45–6
praṇa (pneuma), ix
pregnancy, test for, 149
Priestley, Joseph, 10, 11, 30, 180
prolactin, 149
'protagon', 84, 89, 200
proteins: of brain, 84, 101; chemistry of, 25, 181–2; denaturation of, 199; enzymes as, 22, 24, 25, 51; in neurones, 99, 101–2, 112–13, 116, 117, 118

211

Index

Index

Index